In Search of Meadowlarks

In Search of Meadowlarks

Birds, Farms, and Food in Harmony with the Land

John M. Marzluff

Yale

UNIVERSITY PRESS

New Haven & London

Published with assistance from the foundation established in
memory of Calvin Chapin of the Class of 1788, Yale College.

Yale University Press books may be purchased in quantity for educational,
business, or promotional use. For information, please e-mail sales.press@yale.edu
(U.S. office) or sales@yaleup.co.uk (U.K. office).

Set in Monotype Bulmer type by Tseng Information Systems, Inc.
Printed in the United States of America.

Library of Congress Control Number: 2019947769
ISBN 978-0-300-23714-6 (hardcover : alk. paper)

A catalogue record for this book is available from the British Library.

This paper meets the requirements of ANSI/NISO z39.48-1992 (Permanence of Paper).

10 9 8 7 6 5 4 3 2 1

*For farmers and ranchers who live in harmony with their land
simply because it is the right thing to do*

Contents

Preface

AS THE WORLD STRUGGLES WITH sustaining intensive, industrial-scale agriculture, a wave of mostly young new farmers is hitting the land. Rather than intensifying production to limit the spread of agriculture, they work the land less thoroughly and in concert with supporting natural processes. This new agriculture seeks to provide safe food in a way that also sustains rural lives and the natural biological diversity that shares our planet. New farmers are confident they can feed the world's increasing population by supplying nearby residents a diverse harvest of plants and low-impact meats that wastes less and naturally maintains soil fertility.

The methods the new farmers use go by several names. Some call their approach ecological agriculture or agroecology. To others, it is known as sustainable, restoration, or conservation agriculture. Whatever the name, the practices are in stark contrast to large-scale, conventional agriculture, where pesticides and herbicides made in distant factories are used to grow vast monocultures of corn, wheat, rice, and other grains. Some new farmers employ only organic practices, whereas others seek to sustain their businesses and lands with more traditional methods. The meat and dairy products produced on an ecological farm come not from vast herds crammed into feedlots, where they are sustained by unnatural foods and excessive regimens of pharmaceuticals, but rather from respectfully tended herds. Many ecological farmers step back in time to grow different food plants and animals together and foster a diversity of heritage breeds and crops well suited to their local environment. Such polycultures are more secure than intensive monocultures, because their diversity lowers the risk of crop loss to disease, freak weather, and pests. Perennial crops such as nuts, berries, fungi, and honey are often important diversifying threads in an ecological polyculture, which may be referred to as a permaculture. Ecological farmers work to maintain soil health by eliminating or reducing tilling and composting manure and inedible portions of crops. Returning compost to the soil conserves water and closes the loop connecting a farm's use of soil nutrients with its production of foodstuffs. Restoring what is taken from the local ecosystem sustains continued production without the need for outside supplements. This self-reliance frees the ecological farmer from the demands of big chemical, seed, or stock companies.

My hope for the future is stoked by the new ecological farming movement that is sweeping the land and by new technologies that allow industrial farms to produce more food with less waste and pollution. In both approaches, the nexus between food production and conservation of Earth's biological riches interests and drives me to write this book. In the following pages I recount some of the details that are emerging as to how we might shift diets, improve farm productivity, and more equitably and justly share the bounty, but I do so primarily as they relate to conservation of land for the wild animals that roam in and around our farms. My frame of reference for this consideration comes from the library and the field. There is burgeoning scientific literature that assesses the costs and benefits of coupling intensive agriculture with natural area reservation (so-called land sparing) and contrasts these with a less intensive, wildlife-friendly approach that incorporates elements of nature into farming (so-called land sharing). I am eager to learn how these approaches work on lands I regularly visit.

The scientific frameworks of land sharing and land sparing guide my search for agriculture that is in harmony with the land. In my quest, I travel from a family-run corn and soybean farm in southeast Nebraska, where the birds that once enlivened farm life with their beautiful song have declined, to farms and ranches in Montana, California, Washington, and Costa Rica that welcome wildlife. As production intensifies on traditional farms, I'm pleased to find some native animals, but I see little evidence that increased harvest enables other lands to be spared. On farms that share land with wildlife, I find that agriculture can coexist with a wealth of native birds as well as bears, wolves, monkeys, and sloths. Marketing a diversity of crops and diversifying farm income through tourism, recreation, and environmental education sustains the vibrant lands worked by the ecologically minded farmer.

My search convinces me that farmers of the future need not resign themselves to a life disconnected from the wild voices that sing from their fields. Living in harmony with the land will demand much from us all—farmer and consumer. As we visit farms and ranches that produce grains, meat, fruits, and vegetables, I link our dietary choices to the fate of native wildlife. As we come to know some of the farmers and ranchers I've visited, we quickly appreciate their revolutionary approach to working the land. We will meet fantastic pioneers who value bears, trout, warblers, and trogons every bit as much as they appreciate beef, veggies, and cocoa. Many of these innovators are young—not surprising, given that agriculture is an increasingly popular college degree.

My journey has also been personal. My wife's grandfather was a first-generation Nebraska grain farmer who started raising diverse crops and livestock, mostly with horse and human labor. His farm survives to this day because he embraced the selective breeding, mechanical, and chemical revolutions that enabled corn and soybeans to dominate the farmland of the American Middle West. My daughter and nephew typify the more massive revolution I sensed. Both graduated from college with an interest in farming but honed their skills with a series of internships and on-the-job training. Despite their experience, they shun the revolution in work efficiency that my grandfather-in-law embraced. They work the land much as he did at the start of his career, relying on their muscles and innovations to raise a diversity of crops and animals. Watching the next generation of my family return to the farm intrigued me to learn even more and share what I found.

As I have searched for farms and ranches working in harmony with the land, I have discovered what each of us can do to meet the grand challenge of growing food sustainably. This is not an easy lift, because we will be required to incentivize governments and private citizens to shrink or soften existing agricultural lands, to develop technologies that increase crop performance on distant continents where climates are harsh and fickle, to shift our diets away from overconsumption of ruminant (primarily cow, goat, and sheep) meat and dairy products, and to reduce food waste. Adjusting our actions to meet the challenge is the right thing to do because as the environmental scientist and philosopher Aldo Leopold said, it increases the "stability, integrity, and beauty" of our planet. Failing the challenge dooms future generations to a lonely life made more difficult without the services provided by intact ecosystems and less pleasurable from the loss of daily encounters with wild nature.

Acknowledgments

I AM MOST GRATEFUL TO THE FARMERS, ranchers, and scientists who took time from busy schedules to introduce me to their land and research: Andrew Anderson, Hannibal Anderson, Hilary Anderson, Malou Anderson-Ramirez, Kristin Belair, Xeronimo Castañeda, Arabela Guzmán Cerdas, Barbara Clucas, Isidro García Díaz, Juan Carlos Cruz Díaz, Matt Distler, Rachael Eplee, Jack Ewing, Dean Folkvord, Marcos Alonso García Guzmán, Allison Huysman, Carlos Jiménez, Julie Johnson, Matt Johnson, Sara Johnson, Rodd Kelsey, Sara Kross, Michael Lichti, Jeff Marks, Adam McCurdy, Bridget McNassar, Avery Meeker, John Milesnick, Mary Kay Milesnick, Tom Milesnick, Frank Muller, Brooke Osborne, Jennifer de las Angeles Mora Pérez, Noah Perlut, Dre Ramirez, Marilyn Roberts, Dane Saint George, Jorge Salazar, Juan Luis Salazar, Manuel Sánchez, Alice Sanford, Jessica Schlarbaum, Doris Seaton, Luis Solís, Sara Sweet, Kaeli Swift, Robert Tournay, Orlando Vargas, Paola Vargas, Lauren Walker, Javier Zúñiga, and Raquel Zúñiga.

Curt Meine and Richard Knight were quick to provide access to, and discuss, Aldo Leopold's writings about farming. Estella Leopold broadened my appreciation of Aldo by recounting her dad's relationship with corn. Patrick Tobin, Ryan Garrison, and Christopher Ranger introduced me to insect farmers and pests. My current graduate students, Carol Bogezi, Avery Meeker, Loma Pendergraft, and Kaeli Swift, read the manuscript and offered critical advice for improving its readability.

I'm ever thankful to my family. My wife, Colleen, not only tolerated my research and writing absences but also helped with ranch work in Montana and farming in Nebraska. My daughter Zoe, a farmer and student of sustainable agriculture; daughter Danika, who researched agricultural policy; and nephew Jim, who also farms, motivated me to learn more about their passions.

Jean Thomson Black at Yale University Press supported my ideas in every way, from shepherding the manuscript through the review and publication process to introducing me to formative literature. It has been a privilege to work on this fourth project with Jean. Her associates Margaret Otzel and Michael Deneen readied the initial text and images for processing. Laura Jones Dooley greatly improved the flow and wording of the text with her expert copyediting skills.

The University of Washington School of Environmental and Forest Sciences enabled me to pursue my interest in agriculture's effects on wildlife and supported me during this time in part with the James W. Ridgeway Professorship.

ONE

Meadowlarks in Decline

THE ROLLING HILLS OF RICH, BLACK dirt are striped with neat rows of twin green leaves that have just shouldered their way through the soil toward the sun. As I look through the early morning mist that hovers over this Illinois field, I strain to hear the sweet whistle of a meadowlark's song. Finally, the prairie wind carries the familiar tune from my childhood toward me. I spot the plump, yellow-breasted singer with his contrasting black V-neck plumage atop an old fence post about a football field's distance away. As he throws back his head, the beautiful music pours forth. This bird and its melody are iconic of the American grasslands. Streets, businesses, schools, and celebrities appropriate its name. Six of the United States claim this avian as their state bird. In state pageantry, only the cardinal is more popular. Today, however, meadowlark populations are greatly reduced. A century ago, they were the most common bird in Illinois grasslands, but today ubiquitous red-winged blackbirds and European starlings dominate the meadowlarks' former stronghold. And no wonder—more than half of the meadowlarks' once rich prairie home has been simplified into a monoculture. Today as I listen to the lark, the dominant grass for miles is *Zea mays,* known as maize or corn.

I am creeping along a yellow dirt road through an early spring drizzle a couple of hours west of Chicago to see for myself how the changing prairie has affected birds such as the meadowlark. Other ornithologists have reconnoitered my exact path every spring since 1966. Starting on the small wooden bridge that crosses Mill Creek, passing prominent gray farmhouses, woodlots, fence lines, and telephone poles before ending twenty-five miles later just beyond Shannon Road, this is Illinois Route 002 of the North American Breeding Bird Survey. It is one of nearly 3,700 such roadside rambles annually assessed by volunteers across the United States and Canada. Their efforts enable scientists to measure changes in populations of all the birds that breed in the area. The trend in meadowlark sightings on Illinois Route 002 is not good. In 2014, only sixteen of the birds—nine

Previous page: A male western meadowlark greets the day.

eastern meadowlarks and seven western meadowlarks — were counted. This pales in comparison to the heady counts of nearly one hundred western and more than thirty eastern meadowlarks during the late 1960s and early 1970s.

It takes me an hour and a half before I hear my first meadowlark — an eastern — belting forth a high-pitched, if simple *tingaling* from an abandoned grassy field. A mile farther down the road I encounter the second meadowlark in a pasture rimmed by corn and grassy swales. After three hours I'm at the end of the route, and my tally stands at six larks — three each of the eastern and the western species. This is the lowest meadowlark count ever recorded on Illinois Route 002.

The loss of meadowlarks is symbolic of the plight of birds that once regularly greeted farming families across the globe. Worldwide 5,407 species — 62 percent of rare species classified as threatened on the International Union for Conservation of Nature's Red List — are challenged by expanding and intensifying agricultural practices. Only overhunting and overfishing of rare animals threaten a more significant number. Together these and other human assaults on Earth more than halved the population sizes of mammals, birds, fish, amphibians, and other vertebrates between 1970 and 2012. Grassland birds have been especially hard hit.

According to John Faaborg, an Iowa-born professor of ecology at the University of Missouri, "grassland birds have shown the most consistent declines since the inception of the Breeding Bird Survey." Ornithologist Fritz Knopf agrees, noting that grassland birds have shown "steeper, more consistent, and more geographically widespread declines" than any other group of North American birds. The eastern meadowlark is a case in point: it has declined not only on Illinois Route 002 but also throughout its entire North American breeding range. The estimated population has decreased by 71 percent since 1966. The reason for the decline in farmland birds is fundamentally the loss of grass, but the drivers of this loss extend beyond rampant conversion of prairies and other grasslands into chemically amended row crops. We have plowed up turf and significantly hampered the ability of grasslands to regenerate and rehabilitate themselves, because we extinguished the forces that create and maintain them: fire, bison, and prairie dogs.

Grasslands once covered a quarter of the Earth's land. By the end of the twentieth century, humans had removed over a fifth of the world's native grassland, converting most of it to cropland. In North America, it has been even worse. More than half of this continent's one billion acres of grass is gone; most of this loss occurred west of the Mississippi River beginning in 1850 as homesteaders discovered the fertile soils of Iowa, Kansas, and Nebraska. The rate of the "prairie plow up" from 1850 to 1990 averaged 2.6 million acres each year, which alone could not contain the simultaneous 3.1 million acres' annual increase in cropland. The pace of change has not yet abated—7.3 million acres of prairie, pasture, and hayfields were converted to cropland in the United States from 2008 to 2012. Fifty-five million acres of grass were lost from the Great Plains of the United States and Canada from 2009 through 2015. That loss represents 13 percent of the total that remained and equals an area the size of Kansas. The latest surge in cropland is a result of growing personal affluence, because corn and soybeans are fed to beef cattle crowded into industrial feedlots or distilled into biofuels that in part power our insatiable vehicles. People eat less than 3 percent of the corn raised in the United States.

The inability of the meadowlark to live peacefully among the corn mirrors humanity's increasing struggle to live in harmony with the land. Before the mechanization that followed World War II and the ultraproductive miracle crop varieties and fertilizers that boosted yields of cereal grains in the 1960s, most farmers in middle America and elsewhere worked small holdings of diverse crops and livestock. Meadowlarks found refuge in the pastures and fallow grounds that were interspersed among fields of vegetables, grains, and legumes. This refuge was short-lived as technical advancements, economic policies, and shifting cultural norms conspired to favor fewer, bigger, and more intensively managed farms. Expansive monocultures of wheat, corn, soybeans, rice, rye, barley, or oats became the norm. Pesticides, herbicides, and fertilizers were employed to sustain productivity. Natural field edges, woodlands, and seasonal wetlands were plowed, cut, diked, and drained to increase arable land. Pastures were mowed earlier and more often or forced to support larger herds of livestock as corporate owners, who did not live on the property, sought to wring every last bit of energy from the soil. The degree of such intensification correlates strongly with the declines in farmland wildlife across North America and Europe. Modern farmers in North America are now rarely in the company of prairie chickens, bobolinks, and eastern meadowlarks, whereas their compatriots in Europe would be lucky to hear the call of a stone curlew or see the flash of a red-backed shrike on the hunt for grasshoppers and mice in their fields.

The loss of birds and other natural features from farms is in some sense an inevitable consequence of our need to eat. In response to increased demand, farmers doubled the production of wheat, corn, and a few other grains in the 1960s and 1970s. Food production during this Green Revolution increased annually by 32 million metric tons. And because of new crop varieties that could withstand drought and grow in poor soil and greater use of fertilizers, irrigation, and pesticides, these gains came with relatively small (9 percent) increases in the amount of new land that was brought into cultivation. Norman Borlaug, the father of the Green Revolution, claimed that this intensification spared hundreds of millions of acres from the plow. A 2013 study suggests a more modest but impressive savings of between 45 million and 90 million acres. The Green Revolution has been slow to reach parts of the developing world, where land continues to be cleared to feed the population. In tropical regions, for example, 12 million to 45 million acres of forest are removed annually for crops; 27 percent of tropical forests worldwide have already been converted to agricultural fields.

The careful breeding of new crop varieties during the Green Revolution can reduce the amount of land required to feed people, but doing so has required massive inputs of nitrogen and phosphorus fertilizers as well as irrigation. The runoff of fertilizers and drawdown of aquifers from these activities degrades freshwater, marine, and terrestrial ecosystems and propels an agricultural system that emits about a third of all greenhouse gasses that result from human activity. The methane and nitrous oxide produced by cattle and other ruminant livestock drive climate change more than does our burning of fossil fuels for transportation. One might justify this ecological damage if the human population were adequately nourished, but despite a global glut of food—by some estimates 23 percent more than we currently need—one in every ten people is undernourished. Unequal access to the riches of the harvest relates to social and cultural factors such as affordability, governance, and gender.

Ecologists conclude that "cultivating crops and keeping livestock . . . have had greater impacts on the rest of the planet than any other human activity." And this impact is expected to rise sharply in the coming decades. Global human populations are predicted to top nine billion by 2050 and eleven billion by 2100. Diets are likely to include more significant quantities of meat as personal affluence increases, especially in China and the global South. These projections put us in a bit of a pickle. Although few doubt the possibility of feeding eleven billion people, some believe it will come at a steep cost to our planet. Estimates suggest that nourishing the masses is possible but would require a 10 percent to 20 percent increase in cropland and a 70 percent increase in crop production. That equates to farming an additional 575,000 to 1.2 million square miles, yet only 1.5 million square miles of productive land remain on Earth, which must also accommodate the expanding footprint of our cities as well as the ground required to produce bioenergy, wood products, and the like. Some experts gauge that our needs exceed Earth's capacity by some 2.3 million square miles, an area nearly as large as Australia. It is easy to become pessimistic about the future when confronting such statistics, but let's not lose hope so fast.

A wide range of farmers, academics, and politicians are engaged in solving how best to feed a growing human population without destroying the world. Some consider where existing croplands are underperforming and seek to close such yield gaps with new crop varieties, precise allocation of water and fertilizer, increased sharing of knowledge, and enhanced infrastructure to support the farming industry. They propose that such intensification can be done without further environmental damage because it reduces water use, polluted runoff, and demand for clearing natural areas. In addition to this sustainable intensification, others propose that we must also reformulate the human diet, reduce food waste, or govern agricultural land as a global common to justly nourish humanity. Reducing our future consumption of meat and dairy products from grain-fed cattle, goats, and sheep would allow more existing arable land to grow crops fit for human ingestion. At present, a third of arable land grows food for our livestock—

A grassy swale, where forbs eaten by birds grow, kept within an extensive land-scape of corn is an example of land sharing, or a wildlife-friendly farming practice, in the midwestern United States.

some 3.8 billion of which shared the planet with us in 2016. Culling the herd and eating more plants, poultry, pigs, and nontrawled seafood not only reduce the demand for farmland but also reduce greenhouse gas emissions and improve human health.

The challenge of growing agriculture sustainably while shrinking its footprint dramatically can be met if the world adopts the changes I have just outlined. In fact, research suggests that closing yield gaps, halting farmland expansion, increasing cropping efficiency, shifting diets, and reducing waste can double or triple food production while reducing greenhouse gas emissions, pollution from fertilizers and pesticides, water use, and extinction of native plants and animals. Alternatively, continuing to decelerate human population growth is seen by many as the only way in which our future needs can be fulfilled without extinguishing substantial portions of Earth's biodiversity.

As we farm smarter and eat healthier I wonder what the world will look like in a few centuries. Will we spare remaining wildlands for the rare and shy species that depend on reserves? If we do, will our farms become increasingly hostile to native plants and animals, or will we also learn how to better share agricultural lands with wildlife? As we consider sharing the land I find it useful to reflect on familiar animals, especially ones whose declines are hard to perceive, such as the meadowlark. Despite our being able to glimpse this bird across much of North America, its abundance is in free fall. As I listen to farmers and the birds their lands harbor, I hear the benefits of an agrarian life shared with others. On the prairies and dry grasslands of Illinois, it comes in the form of the meadowlark's song in the still morning air.

FARMING — WHICH I CONSIDER BROADLY to mean the domestication and cultivation of wild plants and animals for food, work, or pleasure—did not come quickly to our early ancestors. The lineage of increasingly upright, innovative, and cultured apes from which we evolved remained on the continent of Africa for most of its 5- to 7-million-year-long history. It wasn't until 45,000 to 60,000 years ago that these brainy walkers, who were less closely tied to the forest than their brethren the chimpanzees, began to spread across the world. Although we belong to the only species alive today, fossil remains indicate that our line was previously much more diverse. Some anthropologists recognize as many as twenty-three species, which they call hominins, our immediate ancestors. Most hominins form a grade of increasing stature and brain size through time. A few of the most recent forms were contemporaries and even interbred. Though we *Homo sapiens* are the only survivors, we are at home on every scrap of Earth's land except Antarctica. We traveled to the Moon and are the only ape to cultivate our own food.

A chimp heritage is easy to see within our DNA. A recent comparison of the human genome—the full complement of our DNA—with that of our closest living relatives, the chimpanzee (*Pan troglodytes*) and the bonobo (*Pan paniscus*), reveals that humans share genes with both. The actual sequences of the nucleic acids that comprise DNA are 96 percent similar between humans and chimps. However, when it comes to differences in the amino acids that these sequences encode, humans and chimps are more than 99 percent similar. This latter statistic is important because amino acids combine to form the proteins that affect our development and contribute to our looks and actions. To understand our agrarian nature, however, we must look beyond our DNA into our culture.

Culture can evolve when members of a group copy the novel ac-

Previous page: Border collies serve as herding assistants on the farm and ranch. These dogs' black and white coloration, rounded heads, and playful behavior are part of the typical domestication syndrome.

tions of others and spread the innovation within and across generations. As individuals learn new behaviors, words, beliefs, and rituals from their group members, some such traditions gain in popularity and help to define a group's culture. This rapid and powerful transformative process has bestowed on all three living members of the chimpanzee family many distinctive characters. Our use of particular tools, languages, and styles of dress evolved through this method. The longevity, social lifestyle, and profound brain of our lineage gave us the ability to innovate and rapidly communicate knowledge that shapes our practices and beliefs in ways that DNA alone cannot. This culture acts as a ratchet to push its own complexity and also to affect the evolution of our genes. In so doing we adapt to our environment and our society, evolving culturally and genetically to meet the challenges of our homeland and its inhabitants. The evolution of farming is a potent example.

Farming is one of three or four hallmarks of the human species. Along with language, art, and technology, especially that which led to effective use of fire, this grand accomplishment distinguished us from the other apes. Distinction in the cultivated field did not come quickly. Although modern humans peopled a considerable portion of Earth for the last 200,000 years, agriculture was developed only as the world began to thaw at the end of the last Ice Age. Generations of people isolated in the Near East, Far East, and North America are thought to have tinkered with the seeds, roots, and wild animals they gathered during the last few interglacial periods when the climate was warm and moist and attempts to sow, reap, and herd might have succeeded. However, failures would have been regular as well because the environment frequently turned cold and dry. The challenges of life when climate returned to cold and dry conditions 13,000 years ago — a period referred to as the Younger Dryas — may have spurred life-saving innovations to domesticate plants and animals. However, this period also is thought to have stalled the widespread adoption of a farming life until the climate was consistently warm and wet, which happened about 11,500 years ago.

The ability to succeed at farming hinged in significant part on our ancestors' abilities to domesticate native plants and animals. While early people may have tried to tame many organisms, few become fully domes-

ticated—that is, completely dependent on humans for their survival and
often reproduction. For example, it is estimated that 2,500 plants in 160
taxonomic families have undergone some degree of cultivation by people,
yet only 250 are considered fully domesticated. In cases involving plants
and many animals, there appears to be a fairly regular progression of domestication through several phases. Humans initially work wild stock on
their native range; confining, stewarding, and harvesting or tending those
with favorable traits. Then, in the improvement phase, the frequency of
desirable forms is increased by selection or favoritism. Later, farmers may
cultivate unique varieties that are adapted to local ecological conditions or
cultural preferences. Finally, stocks are deliberately bred to increase yield,
uniformity, and quality. Purposeful breeding often involves hybridizing native stocks, a practice that is known to have occurred more than 11,000 years
ago during the propagation of figs.

The development of corn (or maize) illustrates the full process of domestication. Farmers in Mexico's Central Balsas River Valley began favoring wild *Balsas teosinte*—a native grass that is the closest known relative of
corn—with exposed grains tightly fastened to the seed head 8,700 years
ago. Variety in the strength of seed attachment is controlled by a single gene
(Tga1 in corn), and the trait may thus respond rapidly to preferential cultivation, but because plants that do not easily shatter their seeds also do not
disperse well, they are often rare in natural populations. The challenges of
identifying and favoring singular nonshattering forms of teosinte did not
dissuade early corn farmers. (In fact, the prowess of farmers to notice and
favor such natural variety rapidly increased the frequency of nonshattering
seed heads during the domestication of most grains, including corn, barley,
buckwheat, and wheat.) Favoritism and selection over the next 2,000 years
increased the frequency of the preferred teosinte cultivar resulting in corn
plants with fewer branches, larger seed heads, and tiny, but recognizable
cobs. Continued selection among initial varieties of corn and the spread
of this valuable crop north and south from its ancestral range in the south-
western Mexican highlands from 2,000 to 6,000 years ago resulted in plants
producing larger cobs and variously colored and shaped kernels suited to
new geographies and local preferences. For example, shriveled ears of pod

corn were favored for ceremonial use by Native Americans, while small-kerneled popcorn was selected in Peru and dent-kerneled corn was preferred by people making hominy and masa. Finally, high-yield corn hybrids that produce a single, uniform, and large ear of seeds per plant, which is dependent upon purposeful breeding by people, were developed during the Green Revolution of the mid-twentieth century.

It would be a mistake to conclude that the dependence produced by domestication demonstrates human mastery over another species or that the domestic species does not also benefit from the relationship. Instead, domestic plants and animals benefit from increased survival and often greatly expanded geographic ranges compared to their wild ancestors. The benefits of associating with humans may have pushed some animals actually to initiate the domestication process. The advantages of docile and tame behavior in wolves (*Canis lupus*) that scavenged human refuse, for example, may have been favored in some populations by natural selection. Approachability may have encouraged some people to prefer tolerant wolves actively, or because of increased access to food and shelter, naturally tolerant wolves may have survived and reproduced exceptionally well around early human settlements. The pathway wolves followed to domestication was likely also taken by other commensal species, such as chickens and pigeons. In other cases, altered hunting strategies, such as only taking young males and allowing females to breed unhindered, may have favored docile behavior. This prey pathway is thought to have started the domestication of a variety of species, including sheep, goats, pigs, cows, reindeer, turkey, and vicuña. Some species were more deliberate objects of domestication, including horses, donkeys, camels, silk moths, rabbits, honeybees, hamsters, and guinea fowl.

Regardless of why tolerance was favored, selection for it is known to produce a wide range of features that make tame animals especially appealing to humans. This suite of traits that evolve together as animals are tamed is known as the domestication syndrome. It includes features such as variegated coat color, reduction in tooth, snout, and head size, floppy ears, curly tails, playful behavior, frequent or nonseasonal estrus cycles, and reduced brain size that we recognize today as typical of domestic wolves, aka dogs (*Canis lupus familiaris*). Why such seemingly disparate traits share a common destiny is not fully understood, but a recent hypothesis suggests it is because of slight deficits in embryonic stem cells from one region that are known to develop into or to control later the development of vertebrate morphology, as well as the nervous, pigmentation, and adrenal systems. Even if domestication initially benefits the wild species that become favored crops, livestock, and working animals, over time humans assume a dominant role.

Thanks to two centuries of scholarship ranging from archaeological field investigations to analyses of ancient DNA, we now have a relatively complete picture of our transition from hunter-gatherer to agrarian people and the role of domestication in that process. Looking at the similarities between the genes of hunter-gatherers and those of early farmers across much of the Old World, we now know that farming developed nearly simultaneously in several areas within the Fertile Crescent. This arc of innovation is a region of the Near East running north from Israel and Jordon (the Levant) through Lebanon and Syria into Turkey (Anatolia) and swinging east through Iraq and Iran. After leaving Africa 45,000 to 60,000 years ago, people first settled in the Fertile Crescent 35,000 to 40,000 years ago. Here, Anatolian, Levant, and Iranian people developed farming independently from one another. These farmers descended from local hunter-gatherers and, after a period of intensive collecting and cultivating of wild cereals, quickly domesticated wheat, barley, and possibly rye. Goats and sheep were also trained, after a transitional period of wild herd management on native pastures, at this time, which anthropologists dub The Neolithic Revolution. Other domestications also likely began during the revolution, even if they flourished slightly later, including oats, lentils, chickpeas, fava beans, figs, olives, cattle, and pigs.

Word of farming and its produce quickly flowed out of the Fertile Crescent and into Europe, Asia, and Africa. Levantine farmers spread south into East Africa about 3,000 years ago. Iranian pastoralists, who grew wheat and grazed sheep, spread north to the higher steppe of the Caucasus Mountains and east to South Asia. Anatolian farmers spread west to mainland Europe 8,500 years ago, but news of their crops preceded the farming technology. For example, people ate wheat in Britain 8,000 years ago, but farmers there are not thought to have grown wheat until two millennia later.

Farming was not solely an invention from the Fertile Crescent. Hunter-gatherers from the high steppes of Eurasia were the first humans to domesticate—or accept the advances of—a wild animal, the wolf. Their dogs and the many breeds they have spawned became essential to the agrarian life that followed. It seems that wherever humans lived, they cultivated the rich local flora and fauna. This appears to have occurred in China at the same time, if not slightly earlier than it did in the Fertile Crescent. The path to producing domestic rice—with its hallmark nonshattering seed heads—is mainly well known. By all accounts, it was a slow process. Over the last 500,000 years, wild rice diversified into several forms suited to the regions just south of retreating glaciers. These primitive forms pushed into India and China around 18,000 years ago and were managed by local hunters and gatherers. Selective harvest, stewardship, and choice of plants with favorable traits produced the first domestic (nonshattering) forms 8,400 to 9,000 years ago in China. At this time, at a village in the Yangtze Basin known as the Huxi site, mixtures of wild and domestic rice plants existed in a ditch structure, likely a precursor to more sophisticated field engineering, flood irrigation, and fire management that characterize later rice agriculture. Wild rice continued to be cultivated in India for millennia without domestication. The rise in domestic rice there did not occur until around 2,000 years ago as Chinese forms spread to India. The Chinese also domesticated peaches and pigs 7,600 to 8,000 years ago, chickens 4,000 to 4,500 years ago, and soybeans 3,000 years ago. Early Pacific Islanders domesticated native fruits and roots, including taro (9,000 years ago), mango (9,200 years ago), and coconut (5,000 years ago). In the Americas, early farmers in the east grew various forms of squash around 10,700 years ago, while in addition to corn, Mexican people domesticated pepper (8,500 years ago), potato (5,000 years ago), and cotton (5,500 years ago). Llamas and alpacas were domesticated in South America 5,000 to 6,000 years ago, roughly at the same time as Europeans domesticated the horse. The rapid spread of agricultural know-how from its many centers of innovation is a pattern consistent with cultural, rather than genetic, evolution.

The invention of agriculture has in part defined us as a species because it so profoundly affected our lifestyle. Rather than relying on our
legs for transportation or our weak sense of smell for hunting, we improved
by harnessing the power of domestic animals, chiefly horses and dogs. Instead of wandering in search of the next meal or suitable materials for cloth-
ing, we stayed home and grew our own grains and fibers. Switching to this
new lifestyle was not without considerable hardship. For starters, it took
the average early farmer a long time—an estimated nine hours—to obtain
the necessary food each day. This effort far exceeded the six hours it is expected to have taken hunter-gatherers. Abandoning a diet rich in protein
for one featuring carbohydrates led to malnourishment of ancient farmers,
witnessed by a loss of physical stature, increased tooth decay, and shortened lifespan relative to hunters and gatherers.

Domesticating animals added protein to the agrarian diet, but it also
compromised human health, as livestock and companion animals exposed
people to many new diseases. As we battled these challenges, our immunity and medical practices evolved. Many children today are vaccinated
for measles and pertussis (whooping cough). The forms of these ailments
that infect humans were derived from viruses and bacteria that infested
our domestic pigs, horses, dogs, and sheep. It is also likely that smallpox,
tuberculosis, tapeworms, and malaria were contracted by our relationships
with domestic animals, although in these cases wild animals may have also
played a complicating role.

The substantial hardship of farm life makes one question why foragers
ever settled down so thoroughly. Favorable climate and prior practice made
farming possible, but it is likely that as our numbers grew, game animals became rare, and agriculture, which was beginning to show promise, became
essential. As the human population grew and became sedentary, changes in
the social dynamic may have provided a final push to a widespread agricultural life. The desire to rise above one's peers or express one's domination
over nature is thought to have spurred the domestication of rare foods, such
as figs, that were featured in competitive feasts. Similarly, the aspirations
for power by individuals and eventually nation-states may have led rulers
to coerce peasants into an agricultural lifestyle.

Yale professor of political science James C. Scott notes in *Against the Grain* that "there is no reason why a forager in most environments would shift to agriculture unless forced to by population pressure or some form of coercion." So, although a rise in the human population may have provided an initial impetus for sustaining the agricultural processes those early foragers routinely practiced, coercion may have further shaped and maintained farming within regions. Social leaders, large landowners, and increasingly formal governing bodies that ruled regional groups and early states may have eventually forced a more permanent agricultural life on their minions. Such coercion benefits the ruling elite by providing cheap (often slave) labor, taxable produce, and a steady flow of foodstuffs. Pressure, however, can and often did lead to rebellion, which together with natural and self-imposed changes in climate and soil fertility produced substantial ebbs and flows in the tenure of early agrarian states.

The cultural influence on early agriculture is especially evident in the Western domestication of chickens and rabbits. In both cases, genetic changes in the animals occurred at the same time that religious decrees were dispensed. Fasting and abstinence from certain foods was common in early religions, but which foods and drinks could be consumed during these times of sacrifice has varied. The eating of fish was often permitted during fasts, but the meat of warm-blooded animals was not. However, in the year 600 Pope Gregory the Great decreed that newborn rabbits were suitable fasting fare because, like fish, they were not "meat." In response, the French monks quickly domesticated European rabbits. Today more than two hundred rabbit breeds provide humanity with food, fiber, research subjects, and pets. Four centuries later, crowded urban living and a proclamation by the Benedictine monastic order enabling the faithful to eat the meat of birds and their eggs during fasts significantly increased the frequency of chicken in the European diet. Asian jungle fowl were shaped for thousands of years before the Benedictine Reform, but rapid changes in their genome 900 to 1,200 years ago indicated the hallmarks of true domestication. At this time, the occurrence of a recessive form of the TSHR gene that is rare in jungle fowl increased dramatically in chickens. Today this once rare form of the gene, which regulates growth, metabolism, and timing of reproduction, is the only type found in chickens. Possession of the recessive TSHR gene causes chickens to increase feeding efficiency and growth rate, lay eggs throughout the year, produce larger eggs, reduce their aggressive tendencies, and lose their fear of humans. These traits enabled increased production of fowl as well as the increased suitability of these fowl for urban husbandry.

The evolutionary impact of agrarian life continues to change our genetic makeup. For example, reliance on cow's milk in northern climes is thought to have favored the evolution of genes that direct the production of lactase, an enzyme some of us produce that is essential for digesting milk. Similarly, an increasingly starchy diet has favored more genes that code for amylase, the enzyme needed to break these complex sugars into smaller bits that we can process. While tasty and nutritious, the sugars in our mouths were exploited by *Streptococcus mutans,* a bacterium that facilitates tooth decay by binding sugars to the tooth surface. We met this challenge with another evolutionary response: an increase in genes efficient at producing salivary agglutinin, which reduces the ability of *Streptococcus* to attack our teeth.

Dogs, our earliest domestic associate on the farm, have also experienced changes in their genetic composition in response to our lifestyle. As our salivary enzymes matched an increasingly starchy diet, so too did similar proteins produced in the dog's pancreas. At the advent of agriculture, the genes coding for pancreatic amylase in farm dogs increased on average fivefold in number. This increase came well after the period of initial domestication and has yet to affect dogs that were not domesticated by farmers. For example, dingoes and huskies do not have elevated amylase gene counts.

Rapid evolution in the culture of farming, including changes in seeds, pesticides, herbicides, and fertilizers, has gifted the average farmer an ability to wring huge crops from the land. And as mechanization evolved to handle the excess, fewer hands were needed on the farm, thus providing a workforce able to revolutionize the industry—both on the farm and in the city.

BEFORE WE TOO QUICKLY revel in our agricultural accomplishments, we should note that many other animals also farm. For example, damselfish

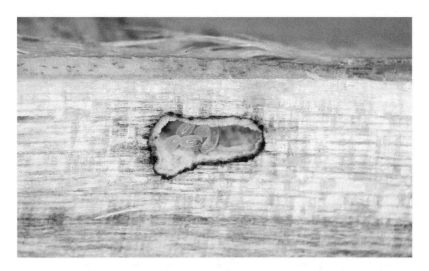

A fungal garden of *Ambrosiella grosmanniae* grows on the walls of a brood gallery created inside a deciduous host tree by the ambrosia beetle *Xylosandrus germanus*. Eggs are laid after the fungal gardens are established; the larvae and adults must then consume the fungal symbiont to properly develop and reproduce. (Photo courtesy of Christopher Ranger)

farm gardens of algae and Yeti crabs raise crops of bacteria. Gardening has also just been discovered in one of our distant relatives: the mouse lemur of Madagascar, a small primate, appears to grow and tend fruiting plants, such as mistletoe. Lemurs enhance mistletoe growth by digesting seeds, which increase their germination. They also cut grooves in gum trees, which facilitate colonization by parasitic mistletoe. As their gardens grow the lemurs seek shelter, especially for their weaned infants, under their dense, rigid branches. While intriguing, the accomplishments of these farmers are dwarfed by those of insects.

Agriculture has evolved independently in three different orders of insects (ants, termites, and ambrosia beetles). These tiny farmers have perfected their craft over 50 million years of practice. Leafcutter ants are a familiar example. Forty species of leafcutter ants harvest pieces of wild plants and transport them to underground farms where they are used as compost to sustain crops of fungi that supply all the ants' dietary needs.

Some 3,400 species of ambrosia beetles do likewise; they raise mushrooms for food and to stifle the defenses that trees employ against them. As the beetle larvae cut galleries within a tree, they feed on cultivated fungi and also use it to block resin and latex exuded by the damaged plant. Fungi-farming insects weed their gardens to remove nonpreferred fungi and use their own secretions and bacteria to control pests and diseases. Success-fully farming a monoculture of mushrooms is not easy, but the insects ac-complish the task by managing a consortium of microbes to fight infection.

Not all insects farm fungi. Black garden ants ranch herds of aphids and harvest the sugary solution they exude. They use physical force and chemistry—biting off the aphids' wings and providing tranquilizers—to slow the movements of their stock. *Melissotarsus* ants are also herders, but unlike the black garden variety, these ants imprison scale insects, known as diaspidids, in burrows under tree bark before eating them. Another ant species, *Philidris nagasau,* from Fiji, has been farming for about 3 million years. These industrious workers gather seeds from six epiphytic plants—those that grow aboveground, typically perched on tree branches in moist tropical forests. Ants gather and then wedge seeds from these plants into cracks in the bark of other trees where they germinate. Ants tend the devel-oping seedlings by fertilizing them with their own waste. In return, as the crop grows, small cavities form in the plants' stems, which provide housing for the ants. These farmers are picky, preferring to harvest and plant only the epiphyte species that produce ant homes.

It seems that other species of farmers have taken to their agrarian lives in much the same way as early humans. However, despite cultivating and tending a wide variety of crops and livestock, nonhuman farmers have maintained a much more balanced relationship with those they have tamed. To my knowledge, humans remain the only species that create new varieties and species of plants and animals through selective breeding and hybrid-ization programs.

THE RISE OF AGRICULTURE not only changed humans and those we tame but also changed Earth. Our croplands and pastures now occupy more than a third of Earth's ice-free land (37 percent, as determined by the World

Bank in 2015). Grazing, plowing, and clearing have simplified the abundant plant life that once carpeted the soil. Fields of a few principal grains and paddocks of pasture grass have replaced 70 percent of Earth's diverse grasslands, 50 percent of savannahs, 45 percent of temperate forests, and 27 percent of tropical forests. Agriculture does not only affect dry land. Farmland irrigation accounts for 70 percent of humanity's use of freshwater. Productive wetlands are drained to increase arable land. The water that still flows through our farms carries inorganic fertilizers — nitrogen and phosphorus, which are often applied too liberally to our crops — to our aquifers, as well as to marine and freshwater ecosystems. There, algae and other plants grow furiously and die, consuming the water's oxygen as they decay and creating dead zones, which wreak havoc on food webs, including many of our most productive fisheries. Although the changes to land and water are visible to the human eye, the invisible effects of agriculture on our atmosphere are equally profound. Real-time assays and analyses of ancient air trapped in the world's ice caps draw a tight link between agriculture and the release of greenhouse gasses that drive climate change. Cropping, which reduced the ability of plants and soil microbes to absorb and store carbon, produced a noticeable rise in atmospheric carbon dioxide as early as 8,000 years ago. Methane also began to increase 5,000 years ago as domestic herds grew. Today, the belches of the nearly four billion ruminant animals (cows, sheep, goats, and buffalo) we herd, as well as the changes in land cover needed to feed them, are responsible for 18 percent of all human-caused greenhouse gas emissions. Those emissions drive climate change more than the exhaust from our planes, trains, and automobiles. Given the substantial effect of agriculture on our land, water, and air, it is no wonder that over 60 percent of species on Earth assessed by the International Union for Conservation of Nature are threatened with extinction by our agrarian life.

As the way we worked the land changed, so too has the wildlife that inhabits farm and field. Farming is indeed a "catalyst for change." Michael Shrubb, a farmer and ornithologist, notes and illustrates this concept by considering how the progression of cultivation has sculpted the English land and in turn favored some birds over others. His story begins 6,000 years ago, when Neolithic farmers began clearing the forests. These early agrarians employed ax, fire, and brute strength to transform the woods into open grassland and heathland over time. Open country birds, such as plovers, larks, and pipits, certainly benefitted from this change. In the mid- to late 1700s, farmers typically grazed their hand-sown croplands after the harvest and then let the field lie fallow for a full year. Cropped fields were sparse, harvest was imperfect, and fallow lands were often plowed. These activities maintained open weedy land, which together with residue from the harvest provided a bounty for quail, sparrows, and finches.

The provision of fallow fields diminished during the nineteenth cen- tury as farmers raised fodder crops for livestock, thus allowing them to inte- grate crop and stock production. Farmers grew a diversity of cereal grains during this time, but rather than let the harvested field lie fallow, farmers replanted it with cover crops, such as clover, and root crops, such as beets, that were suitable winter fare for livestock. This "high farming" practice favored wood pigeons that fed on the tops of fodder beets and corn bunt- ings that ate the waste cereal grain from the field and storage pile. The reli- ance on manure as fertilizer favored a vibrant soil invertebrate fauna, which benefitted thrushes and plovers.

Farmers began to manage their fields with a "three-year-ley" in the late 1800s, which continued for most of the next century. Fallow fields again increased, as fields cropped for three consecutive years were then laid fal- low for three years. This alternation of crops and fallow grassy fields suited farmers who diversified their produce by tending small herds of dairy cows. World economics reduced the acreage farmed in the early 1900s, but this respite from the plow was short-lived as production geared up to feed a country at war from 1939 to 1945. Following the war, farmers increasingly relied on machines and chemicals to boost yields. Tractors quickly replaced horses. Dressings on seeds reduced loss to infection and pests. Birdlife was quick to respond. Seedeaters, such as stock doves, chaffinches, and yellow- hammers, were often poisoned by chemically treated seeds, and the plovers and thrushes that once feasted on worms and other soil invertebrates now found little food in the pest-free soil.

The Green Revolution's hybrid seeds and increasing reliance on chemicals to protect and fertilize crops transformed farming throughout the latter half of the twentieth century, with devastating consequences for biological diversity. Man-made nitrogenous fertilizers replaced manure and negated the need to graze or rest fields. This enabled farmers to specialize in a few crops, notably corn or wheat, rather than growing a diversity of plants selected for their ability to maintain soil fertility or provide food for the live-stock, whose manure amended the soil. The increased use of herbicides, especially those that killed weed seeds before they could germinate, reduced the need to till the soil and eliminated many of the weeds whose flowers and seeds were essential to pollinators, doves, sparrows, buntings, partridge, and larks. The loss of native weeds (aka wildflowers) also reduced the availability of native insects and small rodents that drew warblers and raptors to the farm. The verdant fields, once so full of life, became eerily quiet.

At some point, the pendulum had to swing back.

In the beginning, the swing was slight, but now it is a movement—as indicated by an increase in organic or locally sourced produce—in plain sight within nearly any grocery store. Since 1990, sales of organic foods have increased annually by about 20 percent. However, in their book *Organic Futures,* Connor Fitzmaurice and Brian Gareau are quick to point out that this is just the latest phase in a movement that began in the early twentieth century in England. There, farmers believed their agricultural industry was overly reliant on man-made fertilizers and as a result were willing to sacrifice soil health for crop productivity. The minority view that farms were "living systems in need of understanding and nourishing rather than simply fertilizing" spread to other parts of Europe and resonated especially in the United States during the American Dust Bowl of the 1930s.

As organic growing techniques were improved through experimentation, the counterculture movement of the sixties and seventies gave ecologically friendly farming a second, and distinctly anticapitalist, boost. Farmers' markets and food co-ops multiplied across the United States as activists sought ways to disrupt the chemically laden, corporate industrial, agricultural system. Still, organic products were kept out of the mainstream, in part by the political backing of agribusiness. United States Secretary of Agriculture Earl Butz, for example, pushed farmers to "get big or get out" and farm "fencerow to fencerow."

The success of organic agriculture that we see today arose primarily in response to the food scares of the 1980s and 1990s. Organic produce was seen as the perfect alternative to industrial products that were tainted with carcinogenic pesticides and responsible for toxic and deadly outbreaks of botulism, yersiniosis, salmonella, and *E. coli*. The U.S. Organic Foods Production Act of 1990 brought organic foods into the American mainstream by recognizing and regulating their production.

The organic food revolution continues to gain momentum. Sales of produce from certified organic farms in the United States increased 13 percent from 2014 to 2015, reaching a new record of $6.2 billion. There are now more than twelve thousand certified organic farms that worked 4.4 million acres in 2015. That is a more than tripling of organic acreage since 1997. These farms grew crops on just over half of their land holdings and raised stock on the remainder, which was tended as pasture and range. Organic production in 2015 seems like a lot until you realize that, across the United States in that same year, farmers and ranchers worked 912 million acres. In other words, only a half of 1 percent of U.S. farmland is certified organic. Worldwide in 2016, a paltry 1 percent of humanity's farms are considered sensitive to the local and global ecology. The pendulum has moved, but it has a long way still to travel.

Organic farming is not the only change that is pushing the pendulum. Some conventional and organic farms directly enhance wildlife habitat. These wildlife-friendly practices include not draining ponds or wetlands, intermixing elements of the native surroundings, for example shade trees, among the crops to increase the ease with which migratory animals move through farms, planting winter cover crops that provide feed for waterfowl, fostering native plants along field edges that are used by pollinators and nesting birds, and grazing traditional grasslands that harbor species that specialize on short-grass pastures (such as red-billed choughs in Scotland and Sharpe's longclaw in Africa). Governments often incentivize such practices with direct payments to help farmers offset lost productivity. The European Union, for example, provided $2.7 billion per year from 2000 to 2006 to promote wildlife-friendly farming. The loss of productivity concerns some conservationists, who fear that if we share farms with wildlife,

This sand prairie scrub oak natural area in Illinois is an example of sparing land in the American Midwest.

we will need more farms to feed humanity. Rather than share, they advocate coupling intensive farming with strategies to spare equal amounts of land from the plow. Sparing land in reserves may be the only way to save some shy or challenging animals that rarely survive close to humans.

The argument to continue down the path of agricultural intensification hinges on the ability of future farmers to wring more food from each bit of land already in production and a small amount more that will be added to the agricultural land base. This can be done to an unknown extent by relying on future technological enhancements in the careful application of fertilizers and pesticides, on the efficient use of freshwater, and on genetically engineering crops to produce higher yields. Indeed, these advancements, collectively known as precision agriculture, will be part of the solution. But if this path continues to rely on an ever-shrinking cadre of farmers who depend on large corporations serving the needs of shareholders rather than local residents, then increases in food production are likely to come at a high ecological cost unless land sparing is linked forever with intensification. This linkage may be increasingly difficult, as the economic return possible if the reserved area is farmed rises with the possibilities afforded through precise agriculture.

Advocates of both strategies—sparing or sharing land for wildlife—agree that the challenge to conserve Earth's biological diversity and feed an increasing human population will require changes in how we farm and what we eat. As the human population tops nine billion by midcentury and eleven billion by 2100, farmers will be called on to double or triple their already impressive food production. Those favoring intensification suggest that humanity can be fed without destroying the planet if the advancements of the Green Revolution can be extended more fully to farmers in the developing world and if new discoveries are made, such as how to breed perennial cereal crops, perpetuate high-yield hybrids through asexual reproduction, and supercharge plant photosynthesis. Others are convinced that a lighter touch will provide what we need. Perennial crops, for example, can produce more available energy for people than can annual cereal grains, according to Mark Shepard. This experienced polyculturist argues that a well-tended mix of fruits, nuts, berries, mushrooms, honey, and domestic livestock can produce over 5 million calories of human food per acre. Corn, which on today's industrial ag-lands is mostly raised for biofuel and livestock feed, in contrast only produces three million calories per acre for people. An acre of Shepard's polyculture could provide eight people annual allotments of two thousand calories per day. So, in an ideal world, just over a billion acres of polyculture *could* feed a population of nine billion people, while conserving land productivity and sustaining some biological diversity.

The feasibility of agriculture, both intensive cereal cropping and ecological farming, to feed the world in 2050 was assessed in a much more holistic fashion by a team of social ecologists from Austria. Karl-Heinz Erb and colleagues investigated the ability of five hundred agricultural schemes to provide for the expected 2050 human population while also not encroaching on the world's forests. The schemes were all possible combinations of variation in crop yields, cropland expansion, livestock feeds, human diets, and the origin of livestock products. The research team determined that

60 percent of the schemes were able to supply the human population and maintain the Earth's existing natural and seminatural lands. However, the viability of plans depended mostly on human diet. All options of farming were feasible if humans adopted a vegan diet, and nearly all (94 percent) were possible if humans shifted to a vegetarian diet. Feeding the growing population and maintaining forests is feasible even if that population adopts a typical Western diet, rich in meat, but only if croplands expand greatly into grazing lands and crop yields increase massively. Thus, the trajectory of modern agriculture to increase cereal grain production through intensification is viable, even with little change in diet. Stepping back to a less intensive, lower yield, organic farm model can also feed the world and preserve forests, but only if paired with a vegan or vegetarian diet or if associated with a vast expansion of cropland into existing grazing lands and concomitant intensification of grazing on the remaining pastures.

The feasibility of either traditional intensification or organic diversification of our agricultural systems would be enhanced if we made two additional shifts in our behavior. The first is merely to reduce food waste. In 2013, worldwide we wasted 450 metric tons of food. That was 74 percent higher than only thirty years ago, but the increase was not uniform around the world. In Europe, wastage was cut in half from 1987 to 2013, but in Asia, Africa, and South America wastage is vastly higher now than in the immediate past. Food waste is reduced when shoppers make frequent trips to nearby markets, buying only enough food for a few days. The second shift in our behavior that increases our future agricultural options involves human population growth. Obviously, if future farmers were required to feed fewer people than the projected nine to eleven billion, then their opportunities to meet human needs and provide for biodiversity would be significantly expanded. Therefore, the need to reduce the human growth rate is fundamental to our ability to sustainably harvest the Earth's bounty. A research team led by Eileen Crist from Virginia Tech emphasized this point with a 2017 review of the interaction among population, food, and biodiversity protection. Many countries from diverse cultures have successfully reduced their population growth, mostly by launching well-funded human-rights campaigns that increase the educational opportunities for women and girls, establish accessible and affordable family planning services, provide modern contraceptives, deploy health workers and counselors, mandate sex education in school curricula, and eliminate government subsidies for large families.

OUR FARMS HAVE certainly evolved from their humble beginnings only a few thousand years ago. How we tend these gardens is now central to our sustainable use of Earth. Computer simulations suggest that either intensive management of cereal crops or organic farming of a more diverse agricultural portfolio can feed the world and protect its forests, even as the human population multiplies during the remainder of this century. While computers are useful for such projections, I prefer a firsthand look at our farming options. So, I pitched my tent among crops of beans, grass, and veggies, hiked verdant fields in search of birds and other wildlife, and talked with farmers to see their crops as they do. My adventure begins in the corn belt of Middle America, where today's expansive monocultures give us a glimpse of the future world that opts to intensify its agricultural practices.

THREE
Not That Much

GLIDING ABOVE NEAT ROWS OF SOYBEANS in the cushioned cab of a John Deere combine, Tim Simone is excited about the annual harvest. He closely monitors the bright green craft's position using an onboard global positioning system coupled to a detailed map of the field that tracks each gentle rise and fall of the terrain as well as what has and what remains to be harvested. Simultaneously, he keeps an eye on the moisture content of the beans that are picked. On this hot and clear September day, the seeds have less than 13 percent water content—low enough to keep them from splitting apart as they are threshed from the plant stem, husked from the pod, and added to the growing mass behind our seats in the combine's holding bin. Despite the heat outside, we are comfortable in an air-conditioned cab. Tim explains to me that the land to him is like a child and today he's finally getting to see what it's like as a grown-up. His love for the land and the farming lifestyle is genuine, though squarely utilitarian.

We've cleaned about a mile of beans off the land when a black-tailed jackrabbit gets my attention. It is racing just ahead of the cutter as we gobble up the last few yards of the lagomorph's dining room. As it breaks into the open, Tim notes that hawks, which habitually follow the harvester, pick off most rabbits. However, today the jack is lucky; there are no hawks aloft, and it safely dives into the patch of grass at the field's edge. I am glad this rabbit found a bit of refuge, but I feel like the Once-ler, the character who cut down all the trees in Dr. Seuss's book *The Lorax.* In listening to Tim, I hear enough respect for the soil and its necessities to believe that this farmer, unlike the Once-ler, honestly cares about the sustainability of his fields and their ability to support his family.

As the combine fills with beans, Tim's boss, Sam Olter, and his ten-year-old grandson, Luke, pull up alongside us with the grain wagon. Without breaking stride, the load is augered from the harvester to the wagon. After connecting a few more times, they head to the grain elevator with fif-

Previous page: Little habitat exists within the rows of corn that cover much of the American Midwest.

Looking out from the cab of the combine, we prepare to glean the last rows of soybeans from the field.

teen hundred pounds of fresh soybeans—just one of many back-and-forth sallies they'll drive this day. The efficiency of the harvest is impressive—four grain wagons are in continuous motion sidling up to take loads every mile or so from the pair of combines that run through day and night as long as the beans are suitably dry. When it takes three or four gallons of fuel to power each mile the combine travels, efficiency is not a luxury; it is an economic necessity.

The reckoning of row crop farming occurs during four or five weeks when conditions are right for harvest. The prior costs of insurance, seed, fertilizer, irrigation, and tending the crop are now being repaid in ripe grain. When the team of seven or eight men is humming, they can glean several hundred acres each day. However, here nature is the banker, and it can stall payment with a slight rise in humidity or stop it altogether with a sudden, damaging storm. The risk and reward of the harvest energize the farmers that, to my eyes, are running a mechanical marathon. Like fans at a race, their families come to the fields when school or city work is done to provide

moral support and sustenance during short meal breaks. When conditions
for running deteriorate, the workers huddle up at the farm office to tend
the machines, watch the weather, and plan for the next phase of the race.

Young Luke, who rides shotgun with his grandfather between field
and silo, is the fifth generation of farmers to work this Nebraska soil. His
great-great-grandfather, who moved to Nebraska from Colorado, began
farming eighty acres in 1930, just as American economics plunged into depression and the soil was stripped from the land in great dust storms. Times
were tough, and the Olters were poor, but they held on to the farm when
many others did not. Sam's father purchased the gas station in a nearby
town with his brother in 1954 but returned to the farm in the early 1960s. In
1977, Sam and his brother joined their dad on the farm. They did not own
enough land to make a living solely by raising and selling grain, so they got
into the hog business. They did well with the pigs until 1998, when, as Sam
puts it, "corporate America took over the hog industry much the way they
had done with the chicken industry." Facing a mandate to produce more
for less, they got out of hogs and into grain full time. The brothers, together
with their other sibling and sons-in-law, today tend 7,500 acres of Nebraska
corn and soybean fields. Federal subsidies, some for conservation but most
for grain totaling over $2.8 million from 1995 to 2014, helped offset the sub-
stantial expense of expanding harvest and storage capacity. Surviving as an
agricultural family through the Great Depression and Dust Bowl is a testament to the economic and social sustainability of an adaptable business,
but as an ecologist, I'm also interested in seeing how the native prairie life
has fared.

I PITCH MY TENT between the green and golden rows of soybeans just beyond the family farm's grain silos. Last year this field grew corn: the farmers
rotate crops to enhance soil fertility and stay a step ahead of disease and
pests. With an hour or so before dark, I saddle up one of the farm's ATVs
and tool around looking for wildlife. Next to a small pond, I watch the grass
come alive with frogs. They appear to be Plains leopard frogs, a typical resident here in Fillmore County. Each frog has a prominent creamy line that
extends from just behind each eye to its lower back before breaking up

into blotches. They are snaring meaty grasshoppers with little effort and pay no attention to my intrusion. One large frog slams its long, pink, sticky tongue on a hopper just before it escapes under my foot. On the roadway a common garter snake has been smashed, an unfortunately regular occurrence for reptiles seeking the heat stored within the ribbon of blacktop. As I return to the farm office, a Woodhouse's toad is hunting under the driveway lights for bugs. Pink thunderheads build in the east as the sun slips below the western horizon. Just before dark an American kestrel, the region's smallest falcon, dives from the silo's guy wire into the thick safety of the cedar hedge. None of the animals I've encountered this evening is rare or unexpected, but still, I'm excited about the diversity I've found and anxious to see what the night brings.

Tonight, the bean field has the vibe of a small town's favorite drive-in. The crunching of tiny footsteps surrounds me as mice and rabbits search for fallen beans. My heart jumps, and the rodents freeze as we register the eerie screams of young owls, apparently urging their parents forth, from the cedar thicket. Crickets chorus steadily in the light of a vibrant Milky Way. Asteroids and satellites cut long paths through the slowly rotating night sky. As I drift off, comfortable on the duff of a century of old corn and beans, I hear something more substantial moving outside the tent. I spotlight the field but see nothing. I imagine hungry deer, coursing coyotes, and sneaky raccoons. A few hours later the orange crescent moon drops out of view. Though cozy, sleeping on the farm is far from restful as trucks lumber throughout the night carrying their loads to distant grain elevators.

An autumn wind and a low ceiling of steel-gray clouds have cooled yesterday's corn sweat—the increased temperature caused by transpiring crops that is the new normal for Nebraska in late summer. The sun casts a subtle orange glow in the east as it struggles toward a crack of open sky. A riot of tiny handprints in the gray soil just beyond my tent confirms my suspicion that raccoons were afoot last night. And, as revealed by the single track of heart-shaped hoofprints, at least one deer also pranced through the ripe soy crop. Bracing into the stiff breeze, I hear a high-pitched churring from an invisible flock of birds. Suddenly, the choir—a mixed flock of cliff swallows and eastern kingbirds—surrounds me. These insect specialists

In researching farms and ranches, I often pitched my tent within the fields, as here among the Nebraskan soybeans.

are migrating, expertly harnessing the southern flow of dry polar air that follows the cold front to optimize their southern journey. They encourage me to slam a quick cup of coffee and bird the mix of beans, hedgerows, and corn that are now framed by a violet sky. A lovely Cope's gray treefrog clings to my water bottle, it too not quite ready to let the night pass.

The cedar windbreak is alive with birds and a few fox squirrels. I rack up thirteen species in two hours. Fox sparrows, house wrens, and brown thrashers work the undergrowth. Field sparrows, robins, and mourning doves venture a few yards into the beans. Blue jays, Nashville warblers, and cardinals add brilliance. Red-headed woodpeckers and a great-crested flycatcher sally from bare branches to snap up flying insects, while northern flickers probe the decadence in search of ants. In contrast, the adjacent field of corn, each stalk now heavy with its single, ripe ear of golden grain and towering to a uniform height of nearly eight feet, is virtually devoid of birdlife. A few migrant savannah sparrows and mourning doves flit from the weedy ditches bordering the corn, but only a small flock of red-winged

The northern cardinal, a typical woodland species of the eastern United States, finds a home within the hedgerows of midwestern farms.

blackbirds and one northern flicker alight within the field. I stick mostly to the narrow service roads among the maize. Birding deep in the corn is impossible; visibility is as low as in thick fog, and with leaves rustling in the wind, hearing any flying object quieter than a jet plane isn't realistic. Despite these limitations, I am quick to accept Craig Childs's conclusion that a monoculture of engineered corn is a biological desert. Certainly, relative to the more open and varied fields of beans, the ten-foot-wide hedgerows, and the foot or two of struggling weeds along the pathways I've surveyed, there is little within the sea of corn to offer birds.

As I head back to the farm office for lunch, I am struck by the fact that all the exciting birds I saw this morning were woodland species. Most are adapters that exist in towns in even greater abundance than in this rural setting. The only grassland bird I encountered was the savannah sparrow, which is a visiting migrant to this part of Nebraska. This diminutive avian is one of the continent's most widespread species, breeding from the tundra of northern Alaska to the montane grasslands of Mexico. Where are the meadowlarks, dickcissels, and grasshopper sparrows? I'm thankful not to be carrying a heavy shotgun today because I don't flush a single bobwhite quail or ring-necked pheasant. When I ask Sam about the missing meadowlarks and pheasants, he laments, "You don't see them like you used to." The consensus among the farmers is that low numbers reflect the actions of hawks. I'm thinking that the loss of grass or intensification of farming is to blame. Annual changes along the Breeding Bird Survey routes, such as Route 002 in Illinois where I found so few meadowlarks, offer me a way to test the farmers' ideas.

Nebraskans volunteer to drive more than a thousand miles along forty-six routes counting the state's birds. Calculating annual trends in a bird's abundance within a particular place, such as Nebraska, in a statistically rigorous manner is a snap on the U.S. government's North American Breeding Bird Survey website. Using it, I quickly confirm what the farmers suspect. Hawks have increased. The uptick in raptors has been modest but steady. For example, Swainson's hawks, which frequent open farm and ranch country, increased on average 1.6 percent per year from 1967 to 2013, while the more cosmopolitan red-tailed hawks increased by 4.1 percent over the same period. The increase in predatory birds likely reflects a reduction in the use of toxins, such as DDT, that crippled raptor numbers until banned in the early 1970s, and a general increase in woodlands around human settlements favored by red-tails.

A close look at the Breeding Bird Survey results for Nebraska also confirms the widespread loss of birds I found missing from the fields of corn and beans. Throughout the state, game birds have steadily declined. Pheasants are down 2.2 percent per year, while northern bobwhite quail are down just over 1.5 percent. Prairie songbirds have also declined each year. For example, western meadowlarks have slipped by 1.4 percent per annum, eastern meadowlarks by 2.4 percent, grasshopper sparrows by 1.9 percent, and dickcissels by a more modest 0.6 percent.

So, is the increase in hawks really to blame for the decline in grassland birds? Their annual population trajectories are strongly and inversely related, as the farmers opined. But other changes to the land have also occurred during this time, and that confuses the issue. The vast majority—a whopping 92 percent, or more than forty-five million acres—of Nebraska is crop or ranchland. While the amount of cropland has remained relatively steady since an exponential rise during the late 1800s and early 1900s, the acreage in corn and soybeans increased rather steadily from the late 1960s through 2013. This upsurge in corn and beans correlates strongly with the decline in grassland songbirds and game birds. This relationship is nearly identical to that between hawks and other birds. Another piece of evidence shifts the full blame away from hawks. A delicate hawk, the northern harrier, which floats above grasslands and marshes hunting for mice and nesting birds such as grasshopper sparrows and dickcissels, did not increase as songbirds declined. Instead, like the other grassland specialists, the harrier population fell by 1.4 percent per year.

Although the plowing of Nebraska in the nineteenth century surely hammered grassland birds, today it seems to me that the management of cropland is more detrimental to the surviving birds than is the annual removal of the remaining grassland. The use of herbicides, in particular, is expanding at a rapid rate here, as it has in Britain. The U.S. Department of Agriculture's annual census of agriculture started keeping track of the acreage treated with herbicides in 1974. At that time 3.9 million acres of Nebraska was sprayed for weeds. In 2012, nearly every bit of Nebraska cropland, 17.6 of the 21.6 million acres, was sprayed. Sprayed fields produce no weed seeds or insects for birds to eat and provide no grass cover within which to nest. The recent drop in songbirds and game birds may reflect a rise in predators, but it most certainly also represents a sharp upswing in the use of herbicides to control weeds that once enabled farmland birds to live in the scruffy bits beyond the field.

One last observation points to changes in the land as having affected Nebraska's birds. Although bird populations are often influenced by annual changes in the weather, the alternation of wet and dry years seems unrelated to the steady decline in Nebraska's birds. Rather than crash during droughts when cover and food would be rare and then rebound in wet years, Nebraska's birds have never rebounded. This is reminiscent of an observation made around my home in Seattle. There, forest birds decline as new subdivisions gobble up woody lands. And this decline is consistent despite extreme variation in both rainfall and temperature, which are affected by oscillations between El Niño and La Niña weather patterns in the Pacific. In contrast to areas with ongoing forest removal, in nearby reserves bird populations fluctuate with the prevailing weather pattern. It seems that in Nebraska's cropland a similar play is being enacted as grassland and game birds plummet in response to loss of wild and weedy land, despite the reprieve that a run of wet years should provide.

THOUGH INCREASINGLY RARE, a few bits of fallow and forgotten land still support rare animals near the croplands. At the family farm, the narrow verges of grass alongside the one-lane tracks accessing irrigation pumps among the fields held birds and iconic insects. Along one short stretch, I tallied eleven monarch butterflies feeding on brilliant sapphire blooms of blue sage. Milkweed, a group of plants that larval monarchs feed on with gusto and from which they take alkaloid poisons to thwart would-be predators, grow in the roadside ditches. The number of monarchs has dropped steadily across the continent, in large part reflecting the plowing of their prairie homes and the poisoning of the weeds they work for nectar.

Small scraps of land beyond the plow and sprayer are valuable for biological diversity, but to really appreciate what has been lost from the farm, one must visit a larger remnant of the prairie. A few miles from farm headquarters I marvel at the bustle of life. Neighbors are barking ideas at each other. Parents are cropping the fields around their homes, harvesting a diverse bounty of wild seeds and lettuces. Their children are watching and learning. The kids seem timid, but then they surprise me by bursting into rough-and-tumble play. Suddenly an alarm sounds, and all the residents vanish into underground shelters. Although this might be a typical Midwest town's response to an impending storm, the farmers I'm watching are not human. The architects of this rural town are the chunky rodents known as prairie dogs. Hundreds of dogs are scattered about this small reserve of native prairie.

Prairie dogs are ecological engineers. Their handiwork enriches grass-

Prairie dogs keep the vegetation in their towns short so that they can detect predators quickly.

land biological diversity by providing the shelter and short grass that many others require. Burrowing owls nest in the cool shade afforded by abandoned dog burrows. The dogs' grazing enhances the abundance of lark sparrows, horned larks, Brewer's blackbirds, western kingbirds, and other open field species. The short grass within a dog town is a favorite place for upland sandpipers to find food. Decades ago, market hunting nearly extinguished this grassland oddity that seems out of place so far from the beach. Like most rodents, prairie dogs are important prey for a unique diversity of predators — coyotes, kit foxes, ferruginous hawks, and black-footed ferrets regularly dine on dogs.

Prairie dogs need more than grass; they also need a little respect. Their habits are incompatible with the plow, and the burrows are seen as a hazard to livestock. Some fear their role in the spread of disease, such as sylvatic plague, which because of the dogs' social lifestyle has decimated many towns. But their grazing, which is done to afford them a more unobstructed view of approaching predators and reduce the grasses that compete with the leafy greens that they prefer to eat, is what gets them in the most trouble. Ranchers poison them and encourage gunners to shoot them so that more grass will be available for livestock. Several counties leverage fines against landowners who do not control the spread of prairie dogs from their hold-ings, so even the benevolence of those who value the ecological importance and whimsical nature of the dog is limited. The demise of the prairie dog is a clear case of humans' ability to dominate nature and our willingness to do so even when it diminishes the resilience of our homeland. Today 98 percent of the prairie dogs are gone. Small refuges and open minds are this charismatic species's only hope.

The unbroken ground is also essential to another grassland specialist that I fail to find among the rows of corn and soy. Greater prairie chickens occur sporadically in the vicinity of my campsite, but here their presence is closely tied to reserves and a few agricultural lands taken out of produc-tion and planted with a mix of grass and forbs. Prairie chickens are sensi-tive to the balance of grain and grass on the land. They may benefit when small fields of corn, wheat, and soybeans cover up to a quarter of the land, but when cropland consumes more than 37 percent of an area, chickens de-cline. They are especially sensitive to a disturbance near, and plowing of, so-cially important dancing grounds, known as leks. I fear that we have yet to understand how significant the loss of these birds from rural life is. Future farmers are losing contact with one of nature's great ballerinas and one of its most treasured trophies. Plains Indians fashioned their garb and patterned their ceremonial dances after the chicken's attire and courtship rituals. As a teenager, I couldn't wait to get up early and ply the cold dawn to hunt prai-rie chickens with my dad. We'd marvel at large flocks of the birds as they flew over an abandoned Kansas airfield. For all our efforts we bagged only a single bird, but those memories have lasted a lifetime and helped to frame an ethic of conservation. What ethics are formed when a child can no longer experience the thrill of such a chase, regardless of whether with a camera, binoculars, or a gun?

Some farmers set aside land for prairie chickens, prairie dogs, song-birds, and game birds by voluntarily renting property to the Conservation Reserve Program as stipulated in the annual farm bill approved by Con-gress. Since its inception thirty years ago, this program has helped maintain grass and other wildlife-friendly covers on more than twenty-four million acres of the United States. In the Middle West, farmers that participate get assistance to plant their land with native grass, windbreaks and shelterbelts, stream buffers, or the grasses, shrubs, and forbs known to be favored by upland game birds or pollinators. Those that bust up native sod are penalized by reductions in subsidized crop insurance.

Setting aside land under the Conservation Reserve Program appears to have reversed some long-standing declines in prairie bird populations. James Herkert, an ornithologist with the Nature Conservancy, has compared trends in bird abundance across the north-central United States before and after lands were set aside. Grassland specialists, such as the lark sparrow, dickcissel, bobolink, and grasshopper sparrow, showed much improved population trajectories after the Conservation Reserve Program was enacted (1987–2007) relative to before (1966–1986). Pheasants, which as we saw within Nebraska have declined in areas of extensive crop farming, also stabilized across the region after Conservation Reserve Program lands were set aside. Biologists such as Herkert are optimistic over what the Conservation Reserve Program has done for grassland birds but concerned about the program's future. This concern is justified if we consider the economics of farming.

The rental payments from the Conservation Reserve Program make economic sense to landowners trying to farm marginal lands, where crops do poorly. But in places like Fillmore County, Nebraska, where farmers anticipate 70 bushels of soybeans or 240 bushels of corn per acre, the average 2017 Conservation Reserve Program payment of $138 per acre is a tough sell. The math is pretty simple. Typical prices for soybeans and corn are $9.50 and $3.50 per bushel, respectively. So, an acre of soybeans can gross up to $665, while an acre of corn may fetch $840—both are more than four times as great as the return for conservation. The out-of-pocket cost to plant, tend, nourish, insure, and harvest soybeans or corn can reach $450 per acre depending on the vagaries of fuel and government subsidies, such as those that reduce crop insurance premiums. But in most years, the net return on investing in an acre of grain or beans still exceeds the best possible return from an acre that is set aside. And that is without other government aid for crops, such as the ACRE program, which provides an additional $0.20 per bushel of grain or the support of ethanol production that bolsters and stabilizes the grain market. As I talk with Sam and his brothers, it becomes clear that they appreciate natural grassland and its rich biotic community, but it is equally clear that if they are to survive on this land, then they just can't afford to shift cropland to more natural grassland. Government subsidies to raise crops for biofuel or cattle are generous, but those to spare a few acres for conservation are meager.

NOT ONLY THE NATIVE plants and wildlife have declined across the prairies of Middle America; people have also vanished from most places on the land. Social scientist Robert Wuthnow has documented these changes. His research points to a rapidly changing Middle West that has boomed and declined but remained socially and economically vibrant. Farm life had its heyday from 1890 to 1920 when homesteaders owned their own land and produced a wide assortment of dairy, meat, and grain products. Massive grasshopper invasions, drought, and economic depression in the 1930s forced many farm families off their land, and even though prices for farmed goods increased during the war years, by 1950 the number of farms had been halved and the rural population reduced by 80 percent. The exodus was the result of aging farmers who moved into towns to retire, young people who left farms to attend college, and the poorest farmers who abandoned the unproductive land. Those who prospered increased their land holdings by purchasing failed farms. This trend has continued through the present. In Nebraska, for example, the size of farms in 1950 averaged 443 acres, but in 2015, the total number of farms had been cut in half and their average size had grown to 928 acres. The flip side of this rural flight was a steady increase in the number of people that lived in midwestern towns and cities. This restructuring of the regional population, as Wuthnow points out, has been "overwhelmingly successful" in enabling the region to sustain a strong economy and afford its residents a high quality of life. The area as a whole is a place of resilience and adaptation to changes in nature, technology, and the world's economy. However, regional resilience has come at the expense of many small towns that once were vibrant farming centers. I have come to know one of these places through the eyes of third-generation farmers whose land today is tended by Sam and his brothers.

Ong, Nebraska, is in many ways typical of Middle America's smallest rural towns. First platted in 1886, Ong's population peaked in the late 1910s and early 1920s with 350 to 400 residents. A century later only 63 people remained, scattered among a dozen or so houses. Precipitous declines in population are the rule, not the exception, in small midwestern towns; half of all municipalities with 250 to 500 residents in 1910 were more modest in 1950. And those like Ong that were dependent on agriculture and close to larger centers, in this case, Hastings and Lincoln, had a 70 percent or greater chance of shrinking. The cold statistics of population decline underestimate what has truly been lost from places like Ong. Community—the synergistic actions of hard-working and self-sufficient people—has been shattered. The vibrant and caring life that Scandinavian and German immigrants built is gone; no civic clubs, food banks, or assistance groups remain. Today, there are no industries available, and only one business, a family-run nut and snack confectionary, exists. The lone remaining church holds services only once each Sunday. The productive soil that sustains monocultures of crops and a few fortunate families, such as Sam's, has failed Ong.

Ong had a good run. Former and current residents celebrated "100 years of progress" in 1986. They recalled the boom years when trains ferried passengers back and forth to nearby Geneva and hauled stock to the markets of Chicago and Kansas City. The annual autumn harvest festival drew people to Ong from as far away as Missouri. And they remembered the "dirty thirties" when crops failed and tumbleweeds coated with molasses were substituted as fodder to feed each farmer's small herd of dairy cows. It seemed as though Ong had everything anyone could need: lumber, coal, hardware, farm implements, groceries, cigars, candies, a saloon, restaurants, a community building, a creamery, a pool hall, a feed store, an icehouse, a post office, K-12 schools, five churches, a barber shop, a fire department, an undertaker, a veterinarian, doctors, a blacksmith, a harness maker, a mechanic and filling station, and a telephone switchboard office. Ong always had a newspaper, and for several years a local broom factory and the grain elevator employed many townsfolk. Over the years there was a well-supported Lions club, Odd Fellows lodge, Rebekah lodge, woman's club, car club, motorcycle club, and gun club.

To my mother-in-law, Blanche, and her sister, Dorothy Johansson, Ong was just a "picture town" when they were growing up in the 1930s and 1940s. Their eyes brighten as they recall Main Street: a four-block-long boulevard lined with hedges of *Spirea* that bloomed white in spring. Along the road were a bank, two grocery and dry goods stores, a gas station, the lumber yard, a blacksmith's shop, a creamery that took farmers' extra cream for trade, a hardware store, the post office, Dr. Asa's office and drug store, Bush's Restaurant, and Ardella's Café, where you could get the best cheese sandwich for a nickel. This arterial ended in a circle, around a fancy flagpole where the stars and stripes were raised and lowered each day. Side streets led to the icehouse, churches, and schools, including a high school that was built by the Works Progress Administration in 1930. The sidewalk bricks were so plumb that a kid could roller-skate atop them from one end of town to the other.

Farm kids walked to Ong for school each day, and when it was nice outside, larger social get-togethers were common. Each Tuesday evening about a hundred residents and nearby farm families gathered on benches or in cars along Main Street to watch a free outdoor movie. On Saturday night, activities were less structured. After dinner, Blanche, Dorothy, and their older sister, Sophia, roamed town to decide how to spend their weekly five-cent allowance. Their father, Valter, a first-generation American Swedish farmer, talked about the weather and crops with other men in the park, while their mother, Alma, chatted with the women. Most of the time the girls' nickels were surrendered at Bush's Restaurant for a huge scoop of ice cream, piled high on a cone. As the sisters grew up, they cruised the boulevard on foot with their friends, and as the nights darkened, they paired up and dared to sneak away from the hubbub with a boyfriend.

The Johansson farm, like most during the early to mid-twentieth century in Nebraska, was small but diverse. Valter rotated among crops of wheat, corn, barley, oats, milo, and Kavanaugh cane on the level parts of his 160 acres. He raised dairy cows and cattle on the rougher, pastured portions of the farm. Sheds, barns, and silos housed pigs and chickens and stored feed. A large vegetable garden provided fresh produce. Gooseberry bushes lined the driveway, which entered from the county road through a small orchard of cherries, apples, and crab apples. The whole family helped with farm chores. The girls herded the many chickens and pushed grumpy hens off their nests to gather eggs. Alma cooked and baked the raw produce into daily sustenance. What wasn't consumed or sold was canned or stored for the winter. Blanche helped her dad deliver milk and eggs to customers in Ong. Prince, the collie, herded the cows from milking barn to pasture and back each day. The fields were small, so as the cows were moved, they grazed the wild grasses that grew along the irrigation ditches that served double duty as pathways. The productivity of the farm sustained a growing family and farmhands from throughout the area. During the frantic days of the harvest young men would be hired to help pick and haul the crops. Some would come from as far away as sixty-five miles, and while they worked a month or two, they lived with the Johanssons. Nobody went hungry.

The portion of each crop that was not picked or thrashed for grain was used as fodder for the livestock. Blanche and Dorothy remember working each field to gather and neatly pile the crop stalks into pyramids where they dried as picturesque "tee-pees" before they were hauled to the silo or hay maw for storage. Though Valter and his farmhands aggressively fought weeds, when the storms blew, towering rows of tumbleweeds piled on the windward side of any fence.

The Johanssons did not set aside land for nature, but they did plant hedgerows to help slow the wind and hold the soil. These actions and the generally soft touch of a small-scale and diverse farm on the land sustained healthy numbers of game birds and vibrant autumn gatherings of hunters. Hunters came especially from Kansas, where pheasants were rare, to chase ring-necks across the grain fields. The Johanssons' uncle brought up his friends from Wichita. These and others lodged with farmers and enjoyed benefit dinners organized by the ladies' aid society at the Ong Methodist Church. Indeed, hunting added to the rural Nebraska economy, but the prospect of long days in fair chase sustained many well beyond the Cornhusker State boundaries. I knew that thrill as a young hunter growing up in Kansas and relive it daily when I reflect on a picture from 1973. In it, my father beams broadly as he stands beside a Nebraska hedgerow proudly holding two cock pheasants.

The Johansson daughters did not stay on the farm to carry on their father's and grandfather's life work. Valter and Alma were strong advocates of education, and they saw to it that each of their daughters graduated from Ong High School and beyond. Valuing education and the institutions necessary to serve it was a hallmark of rural Middle America. In 1950 Nebraska led the Union in the percentage of adults who completed high school. Dorothy and Sophia went to business school in Hastings, about an hour northwest of Ong. From there they took careers with the FBI and the navy in California. Blanche graduated from Nebraska Wesleyan University and followed her sisters to southern California, where she became a teacher. Though they built lives outside of Nebraska, each sister remained connected to its land. Valter and Alma bequeathed their farm holdings to their daughters, who now hire Sam and his brothers to farm their 320 acres. The farm has survived, but not the town that it once sustained.

Dorothy and Blanche visit the farm and Ong every year or so. They describe Ong now as "sad." Their eyes suggest something even more profound. As I drive Blanche along the crumbling road past the tumbledown buildings, the abandoned high school, and the church stripped of its beautiful stained glass, her gaze is vacant. The relicts before us bear no resemblance to the town she so loved. This physical place brings none of the joy to her that it once did. There is no reason to linger, to consider, or to explore. There is only reason to leave, and so we do. The community of Ong is sustained only as a memory.

NEBRASKA HAS SURPRISED ME. I expected both the social and natural fabric of this once vast grassland to be in tatters. Indeed, I encountered a small town in ruins and native animals on the run from the harvest. The loss of prairie and rural life is to me heartbreaking. Others have pointed to the soil's reduced ability to store carbon and its reliance on massive inputs of nutrients and energy to continually sustain its monocultures of grain and beans as unsustainable. I agree. But the life I encountered was actually greater than I imagined. The center of a cornfield may be apocalyptic, but taking a broader view, I found a wounded landscape that provided support for people and a bit of nature.

Fields of corn and soybeans as far as the eye can see sustain human life and well-being. In just a century the human population of Nebraska has followed the global trend, receding from the farm and small town to become increasingly composed of urbanites. But this transition, though devastating to places such as Ong, has been nothing short of miraculous to the region as a whole. States in Middle America, such as Nebraska, have the lowest rates of unemployment found in the United States. Cities such as Lincoln and Omaha are of moderate size and are consistently ranked among the nation's cities with the best quality of life, as indicated by high income relative to living costs, excellent health benefits, and enjoyable work-life balance. All of these features add up to a society less stressed than is typical in the United States. However, not all midwesterners enjoy such a life. A 2019 study estimates that the reduction in air quality associated with the production and harvest of corn causes 4,300 people to die prematurely each year in the United States. So, even as some measures of the regional economy and lifestyle appear sustainable, others point to its fragility.

Despite the region's success, few independent farmers have been able to remain on their ancestors' farms. Those that have been fortunate have nourished five or more generations from rich Nebraska soil. They have weathered economic and ecological disasters and today make a good living doing what they most enjoy. Sam loves farming, and as he says, "If you are doing what you love, you don't work a day in your life."

Human life and well-being on the former prairie were to my eyes sustained partly at the expense of nature's grandeur. The last of the magnificent wild herds of bison thundered past Ong in 1871. The wolves and ravens that followed them are also gone, but the changes I've noticed are much more recent. In Middle America, as in Britain and continental Europe, farmland supported viable bird populations through the 1970s. As farm size continued to increase, formerly common hedges bordering fields were cut down and many farmsteads were moved or cleared. As a result, the land's diversity was reduced. Buffers along waterways, parks, and reserves maintain at best an impoverished bestiary. Rather than ebbing and flowing with the ordinarily dry and wet weather cycles, today's open country birds find light fare in soils amended with anhydrous ammonia rather than manure or within fields kept weed free with herbicides. Even though few native prairie species have gone extinct in recent times, our luck may be running out as populations of many species reliant on grassland are reduced and put at risk.

My short time with the Nebraska farmers gave me hope for what could be done to allow Middle America to regain its once strong connection to the prairie's biological riches. But to do so will require those who work the land to see and profit from its full potential. The farmers I met love their land, but they now see it principally for its annual promise of a bountiful harvest. As the wildlife scientist and philosopher Aldo Leopold pointed out nearly a century ago, this view is only half of the story. Leopold grew up a few hundred miles east of Ong, Nebraska, in Burlington, Iowa, at the end of the nineteenth century. Farming was very much a family affair at that time, but during his life, he saw drought and bankruptcy wipe the small farmer from the land. Those who remained, Leopold surmised, increasingly viewed their property as a commodity rather than as a community.

I don't blame today's farmer for ranking the land's economic value above its ecological value, but I believe there is room to reshuffle the rankings. And changes in our consumption can help with the transition. In the United States, corn and soybean farmers have little choice but to farm intensively because of the actions that federal government subsidies reward. More than ninety million acres are planted in corn each year. The acreage planted in corn has increased substantially in the past three decades as demand for corn to feed cattle and supply ethanol plants has driven prices up. Therefore, reducing our consumption of grain-fed beef and our use of corn-based biofuel would lessen demand for corn, lowering its cost, and incentivizing farmers to diversify their crop. That could be good for prairie wildlife if incentives were also provided for those who were willing to conserve or restore prairie on some of their lands.

As we saw earlier in the chapter, the federal Conservation Reserve Program has restored native grass to some midwestern land, but it is ineffective in places like southeast Nebraska because it is not economically competitive. Here, most farms would quickly go out of business resting their cropland under the program. Our government needs to give the farmers a fighting chance to fallow and restore some of their lands by raising the rental payments for program lands to equal the return from chemically intensive farming. Unfortunately, the 2018 farm bill reduced rather than increased support to farmers enrolled in the Conservation Reserve Program.

National investment in conservation programs on agricultural lands for the five-year period 2019–2023 is expected to total $29 billion, roughly the same amount that was spent from 2014 to 2018. That substantial investment supports a variety of landowner incentive programs, including a slight raise in the amount of land that will be supported with Conservation Reserve Program incentives (from twenty-four million to twenty-seven million acres). However, to pay for the increase in conserved land, the compensation a farmer obtains per acre enrolled in the program will be reduced, and the government will match farmer investments in seed mixes intended to conserve native birds, wildlife, and insects at a lower rate than it has in the past. Government officials argue that reducing per acre investment in the program keeps farmers working productive lands and directs conservation efforts to the most vulnerable, least productive areas. Moreover, because incentives will go primarily to small practices within a field rather than paying for the rental of a full field, spending will have a wide reach. The rationale is that by paying farmers only for the land they set aside or restore as wetlands, riparian buffers, or filter strips of grass along field edges and in drainage areas the program will enhance conservation and take less land out of production. Research in Iowa suggests that this approach has promise, but limitations.

Incorporating a little bit of prairie in the Iowa cornfield can have surprisingly significant benefits. Bird abundance increases two- to sevenfold and the number of bird species increases two- to fourfold when as little as 10 percent of row crops are replaced with small patches and strips of prairie grasses and flowers. If 20 percent of the field is planted in prairie rather than corn or soybeans, then grassland specialists such as common yellowthroats and dickcissels inhabit and nest within the bits of native habitat. While these small steps help increase the biodiversity of a cornfield in a way that is appealing to Iowans in general and farmers specifically, they do not conserve other grassland species of concern such as Henslow's sparrows or bobolinks. These grassland birds require large prairies. Sharing row crops just a little with birds by restoring prairie around field edges, inserting small strips and patches within fields, and buffering waterways alone isn't sufficient for conservation, but these strategies, which might be tolerable even as commodity prices soar, can augment the value of land spared entirely from farming.

The sustainable solution to grassland conservation in highly productive, privately owned farmlands such as those in the Midwest might include a variety of incentives, some of which encourage mixing prairie into row crops and others of which reward the resting of larger parcels. Regardless, the government should pay farmers a fair price for the land they are willing to reserve. Shifting allocations in the annual farm bill from subsidies that prop up environmentally damaging farming decisions to those favoring environmentally friendly choices would allow Congress to reward farmers without an increase in government spending. A paltry 6.7 percent of the 2018 farm bill goes to conservation actions of any kind. Indeed, there is room within the program's nearly half-trillion-dollar budget to provide equitable compensation for conservation action. I would be surprised if more farmers wouldn't compete to enroll in such a program and prove that they love and respect the land for its value as a community as well as a commodity.

With an incentivized restoration program that allows farmers to benefit from a conservation ethic, those creatures that today are rare and at risk of extinction may be sustained. Many birds that require grasslands for nesting and rearing their broods are disappearing from lands that are farmed from fencerow to fencerow, but they are resilient in agricultural settings that include lands restored and set aside under the Conservation Reserve Program. An investment in the conservation farmer would surely be repaid by a reduction in national and personal sacrifices required under our Endangered Species Act if more grassland birds are legally defined as threatened or endangered. Rather than passing on to future generations an exhausted land, we could pass on a rejuvenated one. That is a message that might resonate with our elected representatives.

Without incentivized conservation, we will continue to see the spiritual and ecological erosion of productive land. In fact, this is rapidly under way along the western edge of the corn belt. From 2006 to 2011, 1.3 million acres of grassland and prairie in Nebraska, North and South Dakota, Minnesota, and Iowa were converted to corn and soy. This rate of habitat loss is comparable to the razing of tropical forests in Brazil, Malaysia, and Indonesia! And much of the damage has occurred in the duck factory of the United States—the rich prairie pothole region where legions of waterfowl breed each summer. The creep of corn into less productive land has occurred because prices have skyrocketed, mainly in response to increased demand for biofuel. Most corn raised in the United States (40 percent) is fermented and distilled into ethanol, which is blended with petroleum-based fuels for our automobiles.

That the range of corn in North America continues to expand in response to market conditions reveals a problem implementing a land sparing/farm intensification strategy to conserve wildlife in agricultural landscapes. Because land use policies do not require or incentivize grassland and prairie reservation that could offset the clearing of new land for intense conventional farming, owners are free to put as much land into production as the market will bear. Rather than balancing conservation and intensification, as the land sparing model assumes, farming in the corn belt is only intensifying. Intensification is possible, even on the marginal fringes of productive farmland for a couple of reasons. First, the industrial corn farming system—which supports the precision technology required to boost crop yields over an area the size of California each year—had the capital to develop new corn varieties that were productive in the short growing season of the northern prairies. Second, the federal government, rather than incentivizing conservation at market value, offered farmers reduced crop insurance premiums and disaster relief programs that mitigated the risk of raising corn in the face of frequent drought. In response, farmers on the edge of the corn belt did what any rational person would do. They converted sustainable pasture, prairie, and grassland to cropland and returned lands originally conserved under the Conservation Reserve Program to corn. They would have been crazy to do otherwise.

Land sparing appears to be a viable strategy only if increasing yields are directly linked to conservation of other lands. Some Costa Rican coffee farmers, for example, are able to intensify production on a part of their farm if they preserve forest beyond the coffee plantation. The Costa Rican government has strict policies that discourage reduction in forest cover. To meet these requirements and certify their coffee as bird-friendly, many farmers retain and restore trees throughout their plantations that shade coffee. I'll discuss more fully how birds respond to this strategy in chapter 7, but here let's focus on payoffs to the farmers. Shading reduces coffee yields substantially relative to plants grown in full sun. Rather than sprinkle trees through their plantation, some farmers are permitted to raise full-sun coffee if they spare an equal-sized forest. These farms are typically small (less than 12 acres) and owned by single families. By increasing yield on half the farm and sparing the rest, these farmers produced two and a half times as much coffee as they could have by shading their crop. This increased yield more than compensated for the reduction in cropland. Native wildlife also benefitted from the small forest reserves restored on former agricultural land. Bird diversity increased 14 percent in the reserves versus conventional shaded coffee farms, and rare forest-dependent birds especially benefitted from the spared lands. The small-scale nature of the farms, the rapid reforestation of saved land, and the varied landscape that included substantial vast, government-owned reserves, as well as the economics of coffee production, enabled land sparing to work in this situation.

If policies in the United States continue to compensate farmers who sacrifice production for conservation inadequately, then the harvest we reap will be borne on the backs of a once vibrant prairie and farmstead life. Barren autumn fields will draw few young hunters from their suburban homes. The rare remaining grouse that dance or cranes that gather along Nebraska's Platte River may entice some to connect with nature, but the everyday experience with prairie wildlife will fade from memory, and many midwesterners will no longer know the land that has shaped their being.

It doesn't have to be like that.

Jon Foley, executive director of the California Academy of Sciences, suggests that diversifying the crops grown in the corn belt would feed more people, employ more farmers, and increase the resiliency and sustainability of our food system. Rather than betting on an inefficient agricultural system that converts corn into fuel and feed for livestock, Foley argues that we should retool the system into one that grows food for people. Today an acre of corn, which could provide the calories to sustain fourteen people (fifteen million calories/year) only supports three, mainly with dairy and grain-fed pork and beef. Stopping subsidies for fuel and feed corn and instead subsidizing the risks taken by farmers to raise edible grains, oil crops, perennial fruits and nuts, vegetables, and grass-fed livestock would result in more healthy and nutritious foods while preserving rich soil, clean water, and thriving communities for future generations. In this system, I can imagine organic farmers sharing the land with wildlife and conventional farmers sparing marginal lands from production. A resulting mosaic of cropland, grazing land, prairie, and wetland would feed more people than today's endless sea of corn does, and it would allow midwesterners to rediscover their natural heritage.

I find myself thinking of Aldo Leopold as I wander through the most productive land in America. As a student back east at Yale University in the early 1900s, Leopold was anxious for the Christmas break. He often thought of hunting ducks and other game along his favorite sloughs of the Mississippi. However, when he returned home, he was saddened to find that with federal assistance all of his bottomland hunting spots had been diked, drained, and covered with a "blanket of cornstalks." Recalling the event several decades later as a faculty member at the University of Wisconsin, Leopold wrote: "I like corn, but not that much." I hear you, professor, I too like corn, but not that much.

I LEFT THE CORN wondering about the certainty with which large-scale farming operations reduced the ecological and social vibrancy of the land. I got my answer in a Montana wheat field and flour mill. There, the ground is rested every third year to recharge the moisture necessary to grow crops, and the nearby town remains picture perfect.

THE THREE FORKS OF THE MISSOURI River—the Jefferson, Madison, and Gallatin—so named by Meriwether Lewis and William Clark over two hundred years ago to honor their sitting president, secretary of state, and secretary of the Treasury, respectively, were running at flood stage as I crossed into Broadwater County in southwestern Montana. Spring rains and warm temperatures that melted snow packed deep on the surrounding mountains contributed to this year's unusually high runoff. The farms along the Missouri's headwaters would struggle with wet fields this spring but gain little immediately from the bounty. Instead, the abundance of water would quickly drain from this rocky land and be used mostly downstream to quench thirsty municipalities, irrigate midwestern farms, and float barges carrying their grains between ports on the Mississippi and to distant international markets. Where I stand on the windswept eastern plain of the Rocky Mountains, some of the meltwater is seeping into the soil and recharging deep underground seas. This groundwater will be tapped during the hot summer to nourish wheat, alfalfa, barley, and peas that are just now emerging from the soggy soil. The problem is that the taps here and elsewhere are running dry. Out of sight, beneath the verdant spring fields, lurks one of our most daunting challenges to sustaining and increasing farm production—the global depletion of groundwater.

Water stored naturally below ground underlies most of Earth's surface and provides a year-round supply of filtered potable and irrigation water to billions of people. By volume, most of the groundwater that we pump from wells is used to wet our crops. However, nearly half of the world's population also relies on groundwater for their primary drinking water. Not surprisingly, as our population has risen, so has our use of groundwater for both drinking and irrigation. Although the quantity is difficult to estimate, experts suggest that humanity currently uses about 6 percent of the annual recharge to the groundwater system.

Our seemingly slight use of the annual groundwater allotment could

Opposite page: Stubble from last year's harvest defines a fallow wheat field.

be sustainable, but here's the rub: even though 80 percent of aquifers are used at rates below their natural rate of recharge, a few are grossly over tapped. The Ogallala aquifer that supports the corn and soy grown in Nebraska, for example, is currently used nine times faster than it recharges. California's Central Valley, which we will explore in chapter 5, also uses its groundwater at a rate six times greater than it is replenished. The most overused groundwaters are those that support farming the arid plains of India, Pakistan, Iran, Saudi Arabia, Egypt, and Mexico. Massive overruns in our use of some aquifers overshadow our sustainable use of others. Globally, we use groundwater at a rate three times greater than it is restored, and 1.7 billion people depend on aquifers that are shrinking rapidly from overexploitation. This unsustainable use of groundwater threatens current global food production, future increases in food production called on by some to sustain our growing population, and water security.

I came to Montana to see for myself how wheat farming is complicit in groundwater depletion. From reading, I knew that wheat crops are especially hard on our groundwater reserves. Worldwide, almost a quarter of groundwater depletion can be traced to wheat farming. In 2010, for example, wheat farmers, especially those in India, China, Pakistan, and Mexico, drained sixteen cubic miles of groundwater more than is annually recharged to irrigate their crops. That overdraft amounts to more than seventeen *trillion* gallons. This overuse is caused by increasing local and worldwide demand for wheat. In India and China, most of the wheat grown on borrowed water is used to feed their citizens, but nearly a quarter of the groundwater depleted in Pakistan and Mexico is used to raise crops that feed their neighbors in Iran and the United States, respectively. I expected a similar story here on the sagebrush plains above the Missouri River that were homesteaded by farmers just a century and a half ago.

Was I ever wrong!

As I drove alongside the neat fields of winter wheat and chickpeas, I saw no sign of irrigation. Missing were the center pivot contraptions I knew from Nebraska that create crop circles visible from space as they rotate around a well and sprinkle groundwater from the Ogallala onto fields of corn. Also absent were the oversized Tinkertoy-like aluminum pipes and wheels that I've watched field hands in central Washington wrestle into position to draw water from irrigation canals to pastures of grass and alfalfa. I soon learned that the method of farming I was exploring was a long-standing sustainable approach tailored to the arid western United States. It is called dryland farming, and rather than drilling wells to irrigate crops it relies on soil and crop management to capture and store the few inches of rain each field absorbs during the wet season (typically the winter west of the Rockies and spring to early summer east of them). The farmers I met were conventional in their use of fertilizers and herbicides, but when it came to water, they were anything but. They built on the knowledge of their ancestors, who also farmed this dry land, to grow food without depleting their underground water stores. This conservation ethic was born from necessity and refined with ingenuity. As second-generation farmer Dean Folkvord put it, "You could drill a well ten thousand feet down and still not come up with a flow of a thousand gallons per minute, which is what you'd need to irrigate." Dryland farming may have been a forced harmony with the land, but I was eager to learn more.

The approach to dryland farming in Montana has evolved over the past century and a half since the first European homesteaders came west through the mountains following the Oregon Trail. At first, farmers worked the moist soils in the river valleys. But as populations grew, spurred on by propaganda "extolling the virtues of fertile land and nutritious crops" from companies eager to develop a reliance on their fledgling cross-country railroads, less desirable lands were settled and farmed. By the early 1900s, annual conferences were held and books were written to promote standard practices in dryland agriculture. While variable from place to place, a cornerstone of dryland farming was the fallowing of land following the harvest of cereal grain, such as wheat. The fallow land was sometimes used as pasture for livestock, but most farmers tilled it frequently to increase water storage and reduce weeds. Fallow lands were typically plowed deeply during the autumn to bury moist topsoil, and the top few inches of soil were frequently pulverized into dust that inhibited evaporation and protected deeper layers from drying. These methods lost favor during the frequent droughts of the early 1900s, especially during those of the 1930s, because tilled lands were stripped of their fertile topsoil by windstorms. Farmers that survived these Dust Bowl years quickly adopted soil conservation strategies promoted by the federal government, notably reduced tilling of fallow land and using portions of the crop that remained after harvest, such as the stubble of wheat, as mulch. These approaches increased soil moisture content, reduced soil loss, and increased crop yields. But no one tactic fit all needs.

In some cases, when cropping is alternated annually with fallowing, soil moisture builds up to problematic levels. Excess water can leach minerals and salts deep into the soil and carry them along impermeable layers until they seep out in surface flows. As this saline water evaporates, it poisons the ground to the extent that crops can no longer be grown. By the late 1970s farming was no longer possible on hundreds of thousands of acres of western land because of saline seep. For this and other reasons, including vagaries of the local and international grain markets, dryland farmers are exploring new crop rotations that minimize damage to their lands while maximizing their productivity.

DALE FOLKVORD BEGAN dryland farming on the bluffs above the Missouri River in 1958. As he worked the land that Lewis and Clark traversed a century and a half earlier, Dale was quick to innovate. He was the first farmer in the area to stop tilling his land and one of the first to use synthetic fertilizer. Cheap fertilizer was abundant after the Second World War as munitions factories retooled to produce soil amendments rich in nitrogen and phosphorus. Dale's wheat was of high quality—it had more protein content than wheat grown in wetter locales and lower elevations—but his yields were low. In contrast to the fifty to seventy bushels of grain per acre harvested from the relatively moist Palouse of eastern Washington and Idaho, Dale was netting only thirty-five to forty bushels per acre. Despite producing high-quality grain, Dale and his son, Dean, who took over farm operations after graduating with a degree in agriculture from Montana State University, believed that they were not getting paid adequately for it. So, in 1993, as Dean told me, "they decided to do something different." Rather than specialize only in farming, the immigrant father and his son branched out into milling and baking. Their gamble paid off a sustainable livelihood and even produced a world record.

Dale and Dean's Wheat Montana Farms brought customers to the farm to enjoy its bounty. They were early adopters of the farm-to-table movement, and this marketing strategy has been wildly popular with locals and the many travelers who drive the interstate highway that bisects the farm. Central to their business is the Deli at Wheat Montana. The deli features goodies freshly baked from the farm's milled flours as well as loaves of all types of bread. These staples are supplemented with the products necessary to serve sandwiches throughout the day and goods from other local enterprises. While you wait for your meal, you can pick up a t-shirt, hat, bowl, or mug emblazoned with the Wheat Montana logo. I've stopped at the deli for two decades, and never have I not seen it busy. Hungry customers line up at the crack of dawn each day to get fresh scones, cinnamon rolls, or a breakfast burrito. In so doing they come to the farm, which, as Dean puts it, "is still the showcase of everything we do."

Adding value to the grains the farm produces has enabled Dean to build his father's enterprise into a real blockbuster. The mill and bakery don't just feed those who come to the deli but also produce thirty-five thousand to forty-five thousand loaves of bread and five thousand five-pound bags of flour each day that are distributed across the western United States, from Arizona to Alaska. The enterprise also helps other local farmers from whom grains are purchased. The importance of milling and baking to the farm is evident when you consider the workforce. Dean helps two other farmers manage the land and the crops on the nine thousand acres that they own. So, in twenty-five years the farm is basically supporting the same number of agrarians that it employed when Dale and Dean alone worked the land. But, in addition to the 2.5 farmers, the mill and deli each support 30 workers, and the bakery supports 130! In total Wheat Montana Farms employs 200 people and does $35 million in annual sales. Their success garnered the farm Montana's awards for Family Business of the Year and Small Business Person of the Year, as well as the national honor in 1999 of Best Managed Farm. The world record came in 1995, when the team turned wheat in the field into a baker's dozen loaves of bread in just eight minutes, thirteen seconds. (As part of the effort they carted a small mill and thirteen microwave ovens into the field!)

As the milling and baking business boomed, Dean continued to adjust his dryland farming techniques. He and his father had used a typical crop rotation schedule that alternated between crop and fallow. Under this scenario, wheat was planted from September to November, depending on the weather, and harvested in August. After harvest, the field was left fallow until the following autumn, when it was again planted with wheat. Moisture accumulated in the soil and weeds during the fallow year, and as crop residues such as stubble and chaff degraded, the organic content and fertility of the soil increased between plantings. Rather than tilling the land, the farmers sprayed the fallow field weeds with herbicide several months before planting and rolled them into the soil, further enhancing organic content. This scheme worked well—crop yields were stable even in droughts, and soil organic matter increased to 2.5 percent—but it meant that each year, as half the land rested, only the other half produced grain. To increase production, Dean is now in the sixth year of a more complex rotation. Rather than plant wheat after the fallowed land has stored water, the farmers plant chickpeas in May and harvest them in August. When the peas are harvested, the land is immediately replanted with wheat, which is harvested the following August before the field is again rested for a year. Under this rotation only a third of the land is rested each year. But chickpeas, which demand more soil moisture than wheat, provide added organic matter and nitrogen to the soil so that longer periods of productivity between fallowing do not erode the land's health. Cycling between crops of chickpeas and wheat also reduces the need for fungicides because crop rotations disrupt soil diseases such as root rot that thrive when a single crop is repeatedly planted.

Dean doesn't mill the chickpeas he grows, so they don't add to the local economy in nearly the same way as does his wheat. But they are a real cash crop. The forty bushels of wheat that an acre of land produces fetch about $240, whereas the twenty-five bushels of chickpeas from that same acre are worth $450. A diverse cropping system not only puts more land into production each year without compromising the soil's ability to store moisture and nutrients, but also helps buffer farmers from the vagaries of the grain market. According to Dean, "We get more for less and treat the land well."

The herbicides and synthetic fertilizers that are used have the potential to run off fields in wet springs and pollute rivers and lakes, but the more I learned about dryland farming, the more anxious I was to see how the wildlife that lived among the cropped and fallowed lands fared. Dean assured me that the pronghorn, deer, and elk, which are abundant here in Montana, gorge themselves on chickpeas, but what about the larks, grouse, and doves?

I BIRDED SLOWLY along the gravel roads between the wheat and chickpeas as a nearly full moon was setting over the Tobacco Root Mountains. Gradually the stubble in the fallow field before me glowed golden as the sun peeked above the Bridger Range. It was then that I heard the long, chittering song, which seemed to drift down to me from on high. The singer was indeed skyward and cutting a broad arc as he sang. His jet-black tail left me no doubt that I was being serenaded by a horned lark. This sparrow-sized bird doesn't actually have horns, but the black feathers that crown its head pinch up into tiny spikes, reminiscent of a bull's horns, above each eye. As he descended, he quieted, and on landing, he quickly disappeared among the eight-inch-tall wheat stubble.

Flight songs occur in a variety of open country birds. Because these aerial displays are energetically expensive, consuming ten to fifteen times more energy than ground-based display, scientists have suggested a variety of reasons that could have favored their evolution. All elevated singers should benefit from having their message heard widely. Rising above the ground, which absorbs sound, increases the distance over which song can be perceived. But there are many other potential benefits to flight displays. Male bobolinks, a member of the blackbird family that we will come to know better in chapter 10, sing in flight to advertise their quality. Males in good condition, as measured by their mass relative to their body size, display longer than those in poor health, and their stamina increases their ability to attract a mate and successfully rear offspring. Common yellowthroats (which I consider in chapter 6) and dusky flycatchers, both of which

An adult male horned lark stands his ground.

live at the interface of open and shrubby country, appear to use flight displays to distract predators, such as hawks or falcons, from their mate, warn their young of danger, or communicate their abilities and knowledge to the predator. Much like the stiff-legged bouncing of deer and antelope, known as stotting, these birds' flight songs may indicate the bird's awareness and aptitude at outmaneuvering the would-be hunter, thereby dissuading a fruitless pursuit by the predator. In horned larks, flight songs such as I observed, which may be belted out from heights of up to 270 yards, are thought to function in courtship, perhaps indicating the male's quality as they do in bobolinks. Larks also sing from rocks and fence posts, but these function more to defend territorial boundaries rather than to impress potential mates.

I've known horned larks since I was a young boy. To me, they always seemed overly common and indicative of degraded lands. They'd fly from gravel roads in western Kansas and skitter from overgrazed pastures I'd frequent in Arizona. But my 2018 research on the wild tundra of Denali National Park in Alaska gave me a new appreciation for the Lark of the

Mountains, as the father of scientific terminology, Carl Linnaeus, dubbed the species. There, the lark places its nest among the lichens and grasses, and both parents hunt insects for their rapidly growing brood of four or five chicks. Despite occasional hail or snow, it takes only twenty days for a clutch of eggs to result in free-flying young, and after two or three weeks of further parental care, the fledglings are ready to strike out on their own. Larks do not overwinter in Denali, but they do just about everywhere else on the North American continent, often forming large flocks and mixing with other grassland birds. They seem equally at home on a hot, dusty prairie as on the frozen tundra. However, though larks are prevalent throughout their wide range, their numbers have been declining in recent times. Loss of fallow fields, such as the ones on Wheat Montana Farms, may be one reason for their decline.

On breeding bird surveys such as the route I followed in Illinois to begin this book, horned larks have declined 2.5 percent each year since 1968. However, they dominate bird communities in corn and soybean fields, especially if those fields are tilled with conventional methods. In Montana, they have dropped at an annual clip of 2.3 percent. Throughout the state, in 2015 only a third as many larks were sighted during surveys as were seen in 1968. The reason for the lark's plight is due to a steady decline in grazed and bare ground as well as row crop stubble. In the eastern part of the lark's range, these habitats disappeared as abandoned agricultural lands reverted to forests. The spread of corn and wheat crops across the center of North America initially favored increases in lark abundance through the late 1950s, according to studies in Illinois. However, from the 1970s onward farmers have planted and harvested crops earlier in the year, which increasingly overlaps the lark's nesting season. This change in agricultural practice is thought to have been important to recent lark declines detected by the Breeding Bird Survey. In Montana, the loss in lark habitat was due not only to an expansion in alfalfa planting but also to the enrollment of marginal lands in the Conservation Reserve Program (see chapter 3). This has benefitted many obligate grassland birds, but as these set-asides mature, they become too dense for horned larks. As mentioned in chapter 3, James Herkert assessed bird population trends before and after the national adoption of the Conservation Reserve Program in 1987. Of all the species he assessed, only common nighthawks and upland sandpipers suffered more than horned larks after implementation of the Conservation Reserve Program. All three of these species share a penchant for open ground. Wheat Montana Farms has put all its former Conservation Reserve Program land into production for economic reasons, but this may have inadvertently played right into the lark's hand. Each year three thousand acres of fallow land provide just what horned larks require.

Horned larks are not the only bird I found abundant on the farm. Vesper sparrows, another species that fared worse after than before the Conservation Reserve Program was implemented, sang with western meadowlarks from the powerlines that paralleled the road. These species may have especially benefitted from the scraps of land entirely left out of production. Meadowlarks, which have declined so precipitously in Middle America, may do better in the arid West because grazing lands are common and federal lands interspersed among farmland provide refuges. Sagebrush on a Bureau of Land Management holding that bordered Wheat Montana Farms seemed especially attractive to the western meadowlarks I noticed. Vesper sparrows mixed with horned larks and mourning doves in the grassy swales that bound this year's wheat crop. As I walked one swale, I noticed native bunch grasses, blue harebells in flower, and bits of sage. The rocky outposts separating the swale from the adjacent field were commanded by singing vesper and savannah sparrows that dove into the grass and sage as I approached. Unlike the continuous cropping I saw in Nebraska, farmers here left the rough land alone, and the birds readily took to it.

Stan Knudsen, one of Dean's farmers, told me that if I wanted to see a real bird party, I should check out a field where they had recently recycled some bread. Stan and his crew grind up bread too old to sell and spread it as compost. "And boy, do the birds love it," Stan suggested. As I neared the breaded field where new chickpea plants were inching out of the soil, the ground erupted with a pageant of black and white birds. A few common ravens and several hundred California gulls were working the bread and probing for insects attracted to the compost. Their guano would further nitrify the soil, increasing its ability to grow peas. It may seem strange

California gulls among sprouting chickpeas finish up bread repurposed as compost.

that gulls would be common so far from the coast, but the California gull is a dryland specialist.

California gulls winter along the Pacific shore like a typical gull, but they breed in the interior west of North America from central Utah to just north of the Great Slave Lake in Canada's Northwest Territories. Those who study the gull note that the bird will "eat just about any food that fits down its throat," including small mammals, fish, birds, garbage, and a host of invertebrates. The gulls' vast diet and flocking habit help limit insect outbreaks, earning them a revered status among early western settlers. In 1848, for example, a flock of California gulls saved the annual crop planted by Mormon pioneers by eating vast quantities of grasshoppers that were devouring the plants. This deed, known as the miracle of the gulls, is commemorated with the golden gull statue in Salt Lake City's Temple Square and earned the gull its status as Utah's state bird.

Typical of the family, California gulls breed as monogamous pairs, though they often divorce and obtain a new partner the year following failed nesting. Successful nesting colonies are situated on predator-free islands within inland lakes. Lowering lake levels in response to droughts often leaves traditional nesting colonies vulnerable to predators such as coyotes, skunks, and foxes. Such losses may explain why the gull in Montana may be declining. Counts on breeding bird survey routes from 1968 to 2015 fell by an average annual rate of 3.32 percent, but wide-ranging colonial birds are not particularly well suited to roadside counts, such as are done with the Breeding Bird Survey. A better way to assess such species is with dedicated searches of potential breeding areas. A statewide water bird survey was done in Montana from 2009 to 2011, during which more than eight thousand breeding pairs of California gulls were observed, suggesting the species is undoubtedly locally abundant. Providing rich feeding areas, such as the recently planted fields at Wheat Montana, benefit this unique gull, though its persistence hinges on safe breeding spots as well. Fortunately, the state of Montana has provided just what this group of gulls needs. At the southern end of Canyon Ferry Reservoir (an impoundment on the Missouri River) about thirty miles north of the farm, the state created a series of ponds and more than three hundred small islands for nesting birds. White pelicans, Caspian terns, and the state's largest colony of California gulls — nearly half of all estimated breeding pairs — inhabit these islands, which have been designated as an Important Bird Area by the National Audubon Society.

I was pleased to find a typical community of agricultural birds using the chickpea, wheat, and fallow fields that comprise Wheat Montana Farms. The necessity to fallow fields every other or every second year after cropping is an effective way that farmers share their lands with native birds. Open country birds or those associated with sparse grasslands, including Brewer's blackbirds, horned larks, vesper sparrows, lark sparrows, savannah sparrows, killdeer, and mourning doves were most common. Birds less closely tied to grassland, such as yellow-headed blackbirds, barn swallows, Swainson's hawks, Eurasian collared doves, and mountain bluebirds, congregated around the few treed homesteads and a cluster of grain elevators, much as did the woodland birds I found abundant around Nebraska corn farms. At a nearby sage-fringed lake, sandhill cranes and Canada geese tended their broods as eastern and western kingbirds hawked abundant insects. Such refuges remain important in arid agricultural settings just as they do in the much more intensively farmed breadbaskets of the world. Likewise, federal lands spared from cropping provide essential refuges for sagebrush specialists, including Brewer's sparrows and sage thrashers, which sang within earshot of the wheat and chickpea fields.

Despite the mix of lands spared from agriculture and cropped in ways that shared resources with birds, I failed to find any of the larger grouse that once inhabited this region. The greater sage-grouse, a federally threatened bird that depends entirely on extensive sagebrush, no longer exists in Broadwater County. Even here, habitat conversion and human disturbance have driven this iconic bird from its traditional lekking grounds. I also found no sharp-tailed grouse, which though resident in the county, require a rich mixture of native perennial grasses and brush. The conversion of native grassland to annual cropland has reduced this species in Montana, though populations remain viable enough in the state to be hunted for sport. The only members of the grouse family I found in the farmland were smaller, less specialized, and nonnative game birds: red-legged (Hungarian) partridge and ring-necked pheasant.

WHEAT IS AN ANCIENT grain first domesticated by farmers in the southern Levant and southeastern Turkey within the fertile crescent approximately 9,000 to 9,500 years ago. Wild forms of the two principal ancestors of today's wheat, emmer and einkorn wheat, were likely harvested much earlier, up to 19,000 years ago. During the domestication process farmers encountered natural mutations and hybrids, some of which were favored, just as with rice, discussed in chapter 2. Whether favored forms were intentionally selected by farmers or were just those that best withstood early harvest is not known, but the power of cultivation on emmer wheat is seen in the genetic record that indicates a rapid burst in evolution over a few thousand years of human use. Emmer mutants with seeds that did not easily shatter from the stem and chaff that readily separated from the seed on threshing quickly came into fashion with early farmers. These old domestic forms hybridized with goat grass around 8,000 years ago, producing varieties with further improved ease of threshing and likely larger overall size and seed numbers, traits that would be of further utility to early farmers. Plants with these characteristics were soon grown in India, central Asia, and Egypt (~6,000 years ago), where an industry centered on baking bread soon developed. Shortly after that domestic emmer wheat spread to western Europe (~5,000 years ago), the British Isles (~3,000 years ago), and China (~2,000 years ago). Wheat is not currently genetically engineered (and therefore it is all certified as non-GMO), though continued selective breeding for desirable traits has further increased kernel size and number. Selection for semi-dwarf wheat varieties greatly enhanced the productivity and durability of wheat, enabling in part Norman Borlaug's Green Revolution of the mid-twentieth century. Shorter wheat was less prone to lodging, a condition where the seed head bends over to the ground and rots, and because less energy was spent on producing leaf and stem, more was available to increase seed production.

From its humble beginning, wheat continues to this day to be the world's largest agricultural commodity. According to the Food and Agriculture Organization of the United Nations, the annual global production of approximately 827 million tons of wheat is grown on more land than any other crop (544 million acres in 2016) and provides 19 percent of all calories consumed by humans worldwide. The celebration of wheat by past cultures remains evident to this day. For example, the Roman goddess of agriculture, Ceres, whom the ancients credited with providing the harvest in return for cultivating the land, was typically portrayed with a bundle of wheat in her hand. Today, a three-story-tall aluminum Art Deco rendition of Ceres with a sheaf of wheat and bag of its grain, sits atop the Chicago Board of Trade Building. I'd like to think that Ceres, who was responsible for the fertility of the land, would have appreciated what I found in the wheat country of Montana.

As I explored the rolling country and met a pioneering lineage of its farmers, I expected to find a story similar to what I experienced on the industrial farms of Middle America. Despite a conventional and rather large-scale farming operation, I discovered that the farming of wheat in the arid West exists in greater harmony with the land and its resident wild crea-

tures than do the corn and soy farms of the Midwest. The most critical difference in farming between these two landscapes that lightens the dryland
farmer's touch on the land is the yearlong fallowing of fields. Fallow fields
provide nesting and foraging habitat for open ground birds and reduce
the need for fertilizer, herbicide, and pesticide. However, these chemicals
are still used in Montana and elsewhere to varying degrees. In some dryland situations, they poison nearby surface and underground water. The
improved harmony I found in Montana arises out of a necessity to build
up water stores in the soil because groundwater aquifers are deep, too deep
to reach for irrigation. A side effect of this necessity is that growing wheat
in the inland West does not deplete groundwater stores as it does in many
other parts of the world.

The innovative spirit of the wheat farmers I met has not only helped
maintain the integrity of their land but has also helped keep their local
community vibrant. Three Forks, Montana, in stark contrast to Ong, Nebraska, has a growing population of just over two thousand people. Most
are white and of middle age, though Latinos constitute an increasingly significant proportion of the populace and the average age of residents is declining. The annual median household income in 2016 of just over $53,000
is slightly below the national average. Crime is low, and the history of the
region is still celebrated in neat museums, hotels, and festivals. It is the kind
of picture town that today lives only in the collective memory of residents
of Ong. In contrast, a longtime resident of the Three Forks area, Sara Johnson, has seen her town improve since the mid-1970s and remain a pleasant
place to live. She attributes much of the town's success to its proximity to
the small city of Bozeman. Many of Sara's neighbors commute to Bozeman
for work and to enjoy the many amenities its vibrant downtown and university provide. As with Three Forks, the rural towns of Middle America that
survive often do so because of their connection to a small city. Ong had no
such relationship that might attract commuters and offered little to the few
farmers who remained in the area.

In addition to opportunities in nearby Bozeman, Three Forks has benefitted from the milling, baking, packaging, and shipping facility at Wheat
Montana Farms. If the Folkvords simply farmed and sold their wheat, only

a few—not a few hundred—people would be locally employed. The grow-ing and substantial industry of the farm and its produce has helped the city of Three Forks proudly live up to its claim as Montana's Best Small Town.

Wheat Montana Farms is secure as an agricultural landscape for the foreseeable future. Though Dean's children are not interested in farming, milling, or baking, the entire nine thousand acres are protected by a con-servation easement. In return for forfeiting the right to subdivide, mine, or otherwise develop the land, its owners receive individual and corporate income tax credits for fifteen years equal to the amount that their sacrifice reduces the land's value, and they benefit from a similar reduction in the overall value of the taxable estate. As Dean struggles with figuring out how to transition his business to the next generation, it's good to know that the larks, gulls, deer, and elk will continue to have access to the mix of cropped and fallow land that they require. Their productivity will surely keep a smile on Ceres's face.

Farming within the limits of the land or, in this case, water showed me that growing a successful agricultural business can coexist in harmony with its surrounding ecological and social environments. But the benefits to the birds of Wheat Montana are mostly accidental, from the farmers' perspec-tives. Sometimes just letting the land rest is all that nature needs.

FIVE

Working Birds

THROUGHOUT ITS WORLDWIDE RANGE the call of the barn owl is described as "simply a long hissing shriek cssssssshhH." However, what this owl lacks in vocal dexterity it makes up for in regularity. As I camped among old-growth Zinfandel grapes in California's Napa Valley, I was jarred awake every thirty minutes by that pure hissing shriek. The bawling owls were busy feeding a quartet of four-week-old owlets who also shrieked, albeit in more subdued tones, each time their parents came to feed them. Had I not been anticipating the owls' ghoulish wailing, I would have bolted from my tent convinced that a slasher lurked nearby. Fortunately, having spent the day learning about the extensive use of owls by the valley's viticulturists, I quickly fell back asleep after the ghostly white owls with monkeylike faces fed their brood and resumed their silent hunting for voles and gophers.

Like many predators around the world, barn owls have declined, especially in intensively farmed lands. The use of agricultural pesticides first limited the owls. Their eggshells thin if the mother owl ingests small mammals and other prey contaminated with organochloride pollutants, such as DDT. This may have reduced British barn owls during the 1950s and 1960s. With the banning of DDT throughout most of the owls' range, this threat has been dramatically reduced. However, potent second-generation anticoagulant rodenticides, such as brodifacoum, are now taking their toll. In Malaysian oil palm plantations, for example, barn owl mortality caused by secondary poisoning skyrockets after rodenticide applications. Poisons continue to threaten owls today, but agricultural intensification is a more pressing concern. Owls in the Netherlands have declined in response to the loss of granaries that once supported large rodent populations and the coalescence of many small farms into fewer, larger ones devoid of nesting habitat. Similarly, owl populations declined precipitously throughout the midwestern United States as farmers converted rodent-rich alfalfa fields

Previous page: A nest box for kestrels and a high perch for hawks to hunt from entice raptors to a vineyard in California's Dunnigan Hills.

into long rows of cereal crops. Barn owls also often collide with vehicles, and this source of mortality has reduced populations where traffic has increased.

Dangers of the road and on the farm should discourage owls in the Napa Valley. As I gaze over a few of the forty-six thousand acres of wine grapes that line the Napa River and extend up the oaken foothills of the Vaca and Mayacamas Mountains, I see little of obvious benefit to a nocturnal predator. Trellises crisscross the land with wires that would seem to give rodents cover and owls little room to maneuver safely. On some ownerships, only bare dirt and vines exist, suggesting meager chow for a hungry owl. And the traffic is intense as nearly seventeen thousand tourists a day drive a web of roads in search of a favorite red, white, or sparkler crafted by the five hundred wineries that comprise the Napa AVA (American Viticultural Area). How is it then that barn owls have held on here in the heart of California's wine country?

Barn owls owe their success in large part to the goodwill of Napa's vineyard owners and managers. In wildlands, owl populations are often constrained by available nesting sites or food. In agricultural areas, owl prey may be superabundant, but nest sites—abandoned buildings, tree hollows, or rocky crevices—are typically rare. However, in the vineyards both food and nests are plentiful. Gophers and voles are relished by owls, but the rodents pester viticulturists by gnawing on young vines and irrigation systems. To enlist owls as agents of pest control, farmers have saturated wine estates with handmade nesting sites. Most are simple wooden boxes—a plywood cube about two feet tall, wide, and deep—perched atop a fifteen-foot-tall metal pole. The parental owls that disrupted my sleep were tending their brood ensconced safely within such a box constructed and installed by the vintners at Tres Sabores over twenty years ago. More than two thousand similar abodes are provided by winegrowers seeking to lower their maintenance costs within the fifty-mile-long valley.

Matt Johnson, a professor at Humboldt State University, and his graduate students work among the tourists and tasting rooms to monitor owls and determine how the surrounding hilly landscape governs the benefit owls provide to winemakers. Matt has been interested in the provision

Professor Matt Johnson removes an adult barn owl from a nest box.

of such ecosystem services for the entirety of his ornithological career. He's studied the degree to which birds reduce insect pests on coffee in Jamaica and Kenya, as well as their pest-removal services on small organic farms in California. As he told me, "I'm interested in flipping the typical question of how farms are good for birds into how birds are good for farmers." After years in distant coffee plantations Matt began to notice the owl boxes closer to home and in 2014 decided it was time to study them. Now in the fourth year of research, he and his students monitor three hundred boxes across sixty-five vineyards.

I met up with Matt and two graduate students, Allison Huysman and Dane Saint George, at Honig Vineyard and Winery to learn more about their work. The students are excited to have just discovered that an owl they tagged last year as a chick has taken up residence in a box at Frog's Leap winery. She is breeding as a one-year-old bird, something not unusual for the species. But they were surprised at the distance she covered during dispersal: some twenty miles separate her natal and breeding nest boxes. This detail is helping the researchers better understand the dynamics of the owl populations and the areas over which they help farmers. Kristin Belair, the winemaker at Honig, is all ears as the students speak. She welcomes their work on the vineyard, noting that "they are looking for things we might not be aware of." To Kristin, the daughter of a biologist, "science helps us find answers." Research made her aware of the loss of natural nesting cavities on vineyards and therefore the need for boxes. To her the return of owls to the valley—though not well studied, few owls were seen in the 1970s and 1980s—helps "provide diversity and balance to the ecosystem." And the benefits of owls on the vineyard are huge, while the costs are minimal. Owls not only help control gophers and voles in Honig's vines but also are a symbol of the winery's commitment to building a sustainable business.

At Honig, we drive the gravel roads and stop every few minutes to check one of six boxes provided for owls on the estate. Dane and Allison ease a miniature camera attached to an expandable painter's pole into each box's opening to see if any owls are home. The first box is empty, but as they peek into the second, it comes alive with a harsh hissing, more reptilian than avian in nature. The noise confirms what we can now see on a small monitor: a mother owl is tending her brood. The chicks are just a few days old, just what Dane is looking for, but too young for Allison's study. Dane is studying how the landscape affects owlet diets, which requires him to document each bit of food delivered by the parental owls to their growing youngsters. To provide what Dane needs, this box will need some modifications. So, Allison carefully plugs the nest box entrance with a cloth sack and Dane positions a ladder—the type used by orchardists that feature a pole leg for self-support—and climbs skyward. On reaching the box, he replaces the cloth sack Allison has pinned against the opening with an old sock, opens the side of the box, and slips his arm into the box to secure the mother owl. She comes out with little fuss and is handed off to Allison and Matt, who whisk her away to the truck to weigh, measure, and band her for future study.

Dane busies himself installing what amounts to a cable TV system into the owl box. He drills a hole to allow a coaxial cable and electrical wire into the box and secures each to a small camera he mounts on the ceiling of the box. The camera points toward the box entrance so that each night as the parents deliver prey to the owlets, Dane gets a video recording of the event. At the base of the pole, the cables from the camera connect to a battery power supply and a tiny recorder. He'll return every ten days to change the battery and media card and in so doing gain a complete picture of the owl family's contribution to the Honig Estate pest reduction program.

The goal of Dane's graduate research is to estimate the volume and type of prey removed by owls. Other studies suggest an average family of six owls eats nearly a thousand rodents each breeding season. In the nearby Central Valley, for example, each nestling owl receives one to three rodents per night from its parents, amounting to more than a hundred rodents per nestling and more than four hundred per brood during their tenure in the nest box. But Dane and Matt want to know if that is accurate for the Napa owls. So far that number seems about right. After one year of research, Dane has learned that the average brood of owls consumes more than a thousand rodents during the eleven-week-long nestling period. Voles dominate the diet. In most cases, parents procure at least twice as many voles as gophers and supplement their chicks' diets with assorted mice, birds, shrews, moles, and the occasional frog. One exceptional night, owls brought thirty-six rodents to their four-week-old brood of six! This mother-lode included sixteen gophers, fifteen voles, and five mice and rats. Dane is looking forward to adding in this year's results and determining "the actual ecosystem service provided to vintners by the owls." But to really know what the owls do for winemakers, he'll also need Allison's results.

Allison will return to this box in a few weeks. The female will still be roosting with her nestlings, but the kids will be old enough so that mom can also forage on their behalf. At that time, Allison and Dane will again sneak up on the box and capture the sleepy parent. Allison will then fit a small backpack carrying a global positioning device onto the mother owl and return her to the box. As she comes and goes each night the track she travels will be recorded, stored aboard the high-tech tag, and automatically passed on to a computer disk below the nest box. When the tag's batteries give out (about two weeks after deployment), Allison will trap the mother one more time to remove the tag, recharge it, and place it on a different owl. This information is improving the team's understanding of how the amount and distribution of uncultivated grassland affects an owl's use of the vineyard. Matt's former student Xeronimo Castañeda learned that owls seek out uncultivated grasslands, but because commuting is costly, they also prefer to hunt near their nest box. Over the breeding season owls that have little (less than 5 percent) grassland within their home range spend between a quarter and three-quarters of their time in their vineyard. Allison is refining this estimate by observing how owls use vineyards after a change in the surrounding landscape.

An unprecedented wildfire destroyed homes and forests in the hills above Napa Valley the year before Allison began to track owls. These fires generally increase the amount of open ground near upland vineyards, which provides a unique opportunity to determine if the owls using areas studied by Xeronimo reduce their use of vineyards after fire provides additional hunting grounds in the nearby hills. The overall effect of uncultivated land on pest removal from vineyards is not solely a function of an owl's hunting behavior; it also depends on how many owls are in the vineyard. From past work, Matt is confident that as grassland near a vineyard increases, so too will that vineyard's owl population. Allison will help clarify this complication so that Matt will be better able to inform vintners about how the surrounding mix of cultivated and uncultivated land affects the delivery of pest-removal services by owls. He should even be able to suggest where on the landscape more grassland might increase owl use of an estate's nest boxes or where an additional box might increase pest removal by attracting another pair of owls.

Matt's current and past graduate students are painting a complete picture of the services provided by owls and the factors that affect the delivery of that service. Each year of research has seen more and more use of the nest boxes by owls. In 2018 about half of the three hundred boxes monitored contained hungry owl families. What a boon to the barn owl population the simple act of erecting nest boxes has been! And the owls are repaying the favor, helping viticulturists reduce their use of costly pesticides and lowering the damage to vines and infrastructure wrought by voles and gophers. If each family of owls consumes a thousand rodents during the summer and does so by spending about half their efforts within vineyards, then an estate such as Honig, which supports three pairs of owls, can thank them for removing about fifteen hundred pests, every year. It is no wonder that farmers throughout the Napa Valley have embraced the barn owl.

I was astonished by the support of barn owls by the agricultural community of Napa. But this isn't the only rural area that welcomes owls. Thousands of nest boxes have been installed in the borderlands encompassing Israel, Jordan, and the Palestinian territories. Here owls reduce rats within wheat, dates, olives, and pomegranates. The success of barn owls at reducing chemical costs and boosting yields of farmers in Israel has motivated farmers from Spain, Argentina, Uruguay, and Cyprus to start nest box programs. And the mystique of the owl is a powerful tool of diplomacy. Politicians have been unable to forge lasting peace in the Middle East, yet Arab and Israeli farmers and scientists work together to study owls, erect nest boxes, and improve crop production. As owl boxes spill across political boundaries and unite people with different cultures and ideologies, they may prove more iconic of peace than the dove and worthy of consideration for a Nobel Prize.

BARN OWLS AREN'T the only birds to benefit from boxes in the Napa Valley. As I lie in my tent listening to the dawn chorus at Tres Sabores vineyard, the earliest singer is an American robin. But within twenty minutes, robins are accompanied by a rambling black-headed grosbeak, cooing mourning doves, and the churring of western bluebirds. The sound of bluebirds gets me up and moving about the estate, for I had been told by owner Julie Johnson that, in addition to her owl box, she had deployed dozens of boxes for smaller birds, such as the charismatic western bluebird. In fact, the bluebird boxes erected here and throughout the Napa Valley are credited with restoring bluebird populations from near annihilation. Indeed, I immediately confront house wrens, tree swallows, and bluebirds aiming to control the small houses that festoon the grape trellises and fence posts throughout Julie's twelve acres of grapes.

Julie installs and maintains bird boxes on her vineyard because, as she tells me, "it's the right thing to do." As an organic farmer, she is concerned with all the animals on her property and their interactions. She is

A male western bluebird hunts insect pests from a trellis in the Tres Sabores vineyard.

not a fan of the Steller's jays that flock to the tables she has set for visitors to sip wine and munch on breadsticks and pistachios. The jays know that drill, and they steal the goodies, even having learned to flip over heavy ceramic cups to obtain the nuts underneath. But she is convinced that the swallows, and especially the bluebirds, pay rent for their use of her boxes by removing insect pests from the vineyard. Those who have measured insect consumption by bluebirds would agree. Each western bluebird parent requires eight-tenths of an ounce of insects per day to meet their individual metabolic needs. A nesting pair of bluebirds would thus remove a pound of insects every ten days from Julie's vineyard. During the three weeks the pair is feeding their nestlings more than four additional pounds of insects are delivered to the growing brood. This service helps make staying organic and shunning harmful pesticides profitable.

Research conducted by Matt and independently by Julie Jedlicka add credence to Julie Johnson's claims. Their studies demonstrated that bluebirds are especially efficient at removing caterpillars from plots of wine grapes. Using an innovative approach, they staked out hundreds of insect larvae in vineyards and recorded their consumption by birds. In his study, Matt and colleague Katherine Howard placed five mealworm beetle larvae on small cards laid on the ground at intervals within the grapes to simulate outbreaks of caterpillars, such as the European grapevine moth, which can ravage the valuable plants. They discovered that birds found and ate larvae from 30 percent to over 90 percent of the patches within six hours. Remote cameras implicated bluebirds as the most common predator. Jedlicka found a similar role for bluebirds. She and her associates documented a doubling in the number of experimental larvae removed from vineyard areas that included nest boxes relative to regions within the same vineyards that did not include nest boxes. (Rather than using mealworms, this study pinned live beet armyworms, a known vineyard pest, to cards.) As in Matt's research, bluebirds were the main predator of larvae. But, Jedlicka's team went further to reveal that while the occurrence of oak habitat near vineyards helps promote pest removal within the grape fields, the provision of nest boxes increased the number of bluebirds tenfold and therefore the total pest-removal service they provide. Pest removal was greatest below boxes used by bluebirds. There, four of five larvae were typically removed within six hours, but in areas without boxes, only one of five larvae were eaten during the same time. The service bluebirds provide to viticulturists increases during the breeding season as young fledglings join their parents to feed among the grapes. By late summer, when seasonal droughts are the norm, the birds move on to the verdant mountains, and the service they provide declines.

While insectivorous birds, such as western bluebirds, eat the larvae of vineyard pests, they also eat insects and spiders that themselves help keep pest numbers in check. Studies have suggested that despite this potential adverse effect, on balance insectivorous birds are beneficial. But many winegrowers in the Napa Valley and beyond aren't taking chances. They don't just put up nest boxes but also manage their vineyards to enhance beneficial insects and reduce ones that can destroy a vine with a single bite. At Tres Sabores, for example, Julie sows a mix of legume and grass seeds to provide cover on the soil between the rows of grapevines. The resulting carpet of fava bean, vetch, Magnus pea, and grass provides nectar and rearing sites for wasps and other insects that parasitize and prey on leafhoppers, mites, and the notorious glassy-winged sharpshooter. Julie's son, Rory, also a wine-maker, tells me that the sharpshooter hosts the bacterium *Xylella fastidiosa* and spreads it to either growing or dormant vines when it feeds on them. Vines so infected die quickly as the bacteria multiply and block the vine's ability to move water and nutrients from the roots to the rest of the plant. Tres Sabores also conserves two-thirds of its land as native oak and conifer woodland and removes nonnative plants that often host sharpshooters, such as vinca, blackberry, and mugwort. Their efforts to affect the composition of bird and insect predators increase the health of their grapes, but they also increase the resilience of the vineyard to the vagaries of weather.

Cover crops under the grapes provide organic matter for Julie and Rory's soil, which along with compost helps build a vigorous microbiota, which they believe "allows their grapes to express their own essence during the process of making wine." Thanks to decades of soil improvement, growing and mulching cover crops, and building water-retention capacity, the family does not irrigate the majority of their vineyard, even during the toughest droughts. Rather, they welcome variety in weather from year to year because this gives each vintage its unique "personality." Julie does not strive for uniformity in taste; rather she embraces a riskier approach that is economically viable while also increasing her soil's health. She and her patrons look forward to tasting what the vines are able to make from the soil as well as the year's conditions. I can vouch for the vineyard's success in this respect: although both the 2014 and 2015 estate Zinfandel vintages were excellent, my less-than-sophisticated wine nose and tongue could easily tell the difference. I noticed that the bouquet of the 2015 offered strong hints of spice, while the 2014 had a most pleasing peppery finish!

FOCUSING ON THE ecological health of the entire farm has enabled both Tres Sabores and Honig to variously certify their land and business through Napa Green, Fish Friendly Farming, California Certified Organic Farmers, and the California Sustainable Winegrowing Alliance. These programs require active stewardship and detailed management plans that are verified by outside inspectors. To be Napa Green, for example, Julie and Kristin not only employ cover crops and set aside native habitat to reduce erosion and support native insects and birds but also recycle, conserve energy and water, minimize waste generation, and pay an equitable wage to combat so-cial injustices in the industry. Vintners and growers must develop dynamic farm plans that promote conservation and sustain the viability of their entire operations. Honig makes extensive use of solar power, actively participates in local Napa River Restoration and habitat enhancement projects, farms to improve soil health, composts grape pomace and stems at an off-site facility, and engages in many other activities to minimize its resource use and carbon footprint. Similarly, Julie seeks to recycle grape stems, seeds, and skins by composting them with the manure from her sheep and guinea fowl. The compost boosts the soil around plantings outside the vineyard, such as pomegranates, buckwheat, olives, and sage, which are used to cre-ate additional products (balsamic vinegar and olive oil) and provide polli-nator habitat for native insects and hummingbirds. At Honig, bees flourish on their pollinator habitat and are kept to make honey. Both winemakers seek to reduce water use in a very water-intensive industry.

Nearly 40 percent of vineyards in the Napa Valley are green certi-fied. Given the difficulty of planning and implementation, this surprised me. Measuring, while time consuming, allows one to assess how resources are allocated fully. It can paint a picture that leads to improved ways of doing things that not only enhance farming systems or increase wine quality but also lower environmental impact. Kristin found that the process makes winegrowers, such as her, ask questions and always look for improved ways of farming and making wine. It is a dynamic rather than static approach that necessarily must be balanced with the end result of growing and pro-ducing wine.

Organic and sustainable farming systems are two different pathways that ultimately share a common goal of lowering the environmental impact of farming and enhancing long-term land stewardship. Organic farming plans for diversity and natural resistance to disease and pest pressures, as well as soil health, and minimizes external inputs. All inputs must be regis-tered as organic in origin and approved for use in each vineyard. Sustain-able systems may use many organic contributions, but they also include many other considerations such as soil health, carbon footprint, and habitat enhancement and preservation. Organic farming may consist of these and other factors, and many organic farmers incorporate the broad view that certified sustainable systems do. Both methods require extensive documentation and long-term commitment. It takes three years to become a certified organic grower.

Maintaining the certification requires on-site inspections by an independent certifying agent. Taking the considerable steps to certify vineyards under any one or more of these programs by setting aside wildlands, planting hedgerows, taking active measures against erosion, monitoring inputs, and encouraging and expanding habitat for native birds, not to mention dedicating the time and financing the commitment, may not be possible in all agricultural ventures. In fact, we will see in chapter 7 that certification is beyond the economic reach of some Costa Rican farmers. But in the Napa Valley, where land is valuable, tourism depends in part on the scenic natural beauty of native land, and wine sells for a premium, following the practices and becoming certified can directly enhance a farm's bottom line. Moreover, vintners such as Honig and Tres Sabores strongly feel that making such an investment is just the right thing to do.

Another dimension of sustainability also flourishes within the Napa Valley. Many vintners and viticulturists are second- and third-generation members of families who pioneered winemaking in the valley. These next gens have their own organization and celebrate the continuance of a special agricultural life. Julie couldn't be happier or prouder that Rory is now making his own wine, Clader, as well as working with her in the vineyard and cellar at Tres Sabores. She sees it as "nice to have a second generation that is taking what we learned and moved it along." And I see that as the key to long-term sustainability. Rory will evolve, as any progressive farmer must, but he will also carry forth the basic land ethic that he has seen his mother model her entire life. As I leave the vineyard, sheep are mowing the driveway verges, lesser goldfinches work the seed heads of grasses among the vines, and a pileated woodpecker advertises his ownership of the upland forest with an emphatic cry. Life here, in so many varied forms, is in harmony, and it is comforting to know it will remain so for many decades to come.

THE DUNNIGAN HILLS lacks the name recognition among wine drinkers that the Napa Valley enjoys, but here too birds are working to reduce vineyard pests. Nest boxes for barn owls, screech owls, and American kestrels rise above the grapevines, just as I saw in Napa. But what catches my eye are the elevated wine barrel staves. The gracefully curved, oak cask segments float above the neat rows of grapes atop ten- to twenty-foot-tall poles, re-creating strangely the trees they once were. Professor Sara Kross explains to me that these perches are to encourage hunting within the vineyards by golden eagles and other large sit-and-wait predators, such as red-tailed hawks. The eagles have mostly flown north to breed when I visit in early April, but the red-tails are abundant. We count over a dozen as we enter the low, rolling hill country noting their variation in plumage—from chocolate-brown melanistic forms to those dressed in more typical whites and earth tones—and use of the man-made perches.

Kross, who teaches classes in agroecology and conservation at nearby Sacramento State University, is an expert in the services that raptors and other birds provide to farmers. Her research among the Tempranillo and Chardonnay vines has demonstrated how hawks and owls complement each other in the Dunnigan Hills. As in Napa, the owls take mice, voles, and gophers. But the owls are too small or forage at the wrong time to help rid vineyards of other pests, such as black-tailed jackrabbits and California ground squirrels. That is where the hawks and eagles come in. The fear they strike in the hearts of ground squirrels and those they consume not only reduces gnawing on the vines but also makes maintenance within the vineyards safer. Where ground squirrels are common and bold, their burrows are numerous. Rocketing over an unseen burrow swings a tractor's implements wildly, which can damage the tightly spaced grapevines.

Sara began her studies of raptors in vineyards a world away from California. As a graduate student at the University of Canterbury in New Zealand, she discovered not only that predators can help farmers but also that farmers can help conserve endangered predators. As with nearly all native birds in New Zealand, its only falcon is threatened with extinction. Approximately four thousand pairs of the New Zealand falcon (*Falco novaeseelandiae*), known locally as the bush hawk or, in Maori, as the kárearea, remain in scattered populations. Its bold brown and white plumage accentuates its bright yellow cere and rusty red thigh feathers. Unlike most birds of prey this falcon nests on the ground, where it is vulnerable to a variety of introduced predators, including feral cats, stoats, and rats. As a conservation strategy, a program known as Falcons for Grapes moves chicks from the highlands to the vineyards around Marlborough, where they have established a small breeding population. Sara took advantage of these reintroductions to measure the effect of farm life on falcons and the impact of falcons on grape damage.

Sara counted bird pests and grape damage in vineyards with and without falcons over two years to precisely measure the benefits falcons provided to farmers. In vineyards with falcons the abundance of nonnative songbirds that remove grapes (song thrush, blackbird, starling) dropped by at least 79 percent, whereas the native silvereye, which pecks grapes, did not significantly decline in the presence of falcons. Reductions in grape-loving songbirds resulted in substantial decreases in grape damage throughout the Sauvignon Blanc and Pinot Noir vineyards that catered to falcons. Vineyards with falcons benefitted by a 95 percent reduction in the removal of grapes and a 55 percent reduction in the pecking of grapes by songbirds. These benefits netted wine producers a savings of $234 on each 2.5 acres of Sauvignon Blanc grapes and $326 on each 2.5 acres of Pinot Noir grapes they raised.

Abundant songbirds and relatively few predators in the vineyards made them successful nesting grounds for Sara Kross's falcons. She and her colleagues assessed over 2,500 hours of video recordings at vineyard and wildland falcon nests to compare the behavior of falcon parents and the influence of their actions on a nest's fate. They discovered that parents in vineyards fed and brooded their young and attended their nests at higher rates than did those in the distant mountains. Although predation on nesting falcons is rare, Sara observed a feral cat destroy the brood at one wildland nest, and other dangerous predators attempt to raid others. At vineyard nests, she did not find dangerous mammalian predators but did note small owls and harriers unsuccessfully trying to pilfer falcon nests. Overall, she concluded that falcons in vineyards were likely to suffer lower rates of nest predation than were those nesting elsewhere. Her experiences in agricultural settings led her to believe that "translocating New Zealand falcons into vineyards has the potential for both the conservation of this species and for providing biological control services to agriculture."

Sara is collaborating with a graduate student from the nearby University of California, Davis, to determine if the benefits derived by vintners in New Zealand might also be accrued by those in the Dunnigan Hills of California. We meet up with the grad, Jessica Schlarbaum, and a covey of undergraduate assistants as they check nest boxes for the presence of a local falcon, the American kestrel. Jessica is just starting her graduate project and is eager to check each of her twenty-five nest boxes. As with Matt Johnson's students, Jessica eases a camera up to the box and watches as it relays an image to her cellphone. We see only a clean wooden floor — no falcons here. The male kestrel screeching from the hedgerow as we approach the next box suggests that we'll have better luck here. Sure thing: a female huddles low on a set of white eggs. The scene is quiet at the next box, but Jessica is thrilled to find it occupied by a tiny screech owl, also incubating her clutch. The results are encouraging, and with time surely more kestrels and owls will take up residence among the vines. As they do, Jessica and Sara will not only watch the raptors but will also walk among the grapes, check for damage, and convey their results to winemakers, making widely known the economic benefits they might reap if they only encourage predators to live and hunt on their land.

FLYING INTO CALIFORNIA's Central Valley is like skimming over a vast sea. The flat terrain, punctured only slightly by the Sutter Buttes, stretches green and brown in all directions. Glistening rectangular rice fields line both sides of the Sacramento River as it chugs through the valley toward the Pacific. Straddling the band of rice are fields that grow over 230 different crops, producing 8 percent of the national agricultural value from 1 percent of its agricultural land. Sara Kross has expanded her studies of raptors in vineyards to understand better how farmers in this vast agrarian sea can benefit by drawing birds near their crops.

Many farmers in the Central Valley manage portfolios of fields that produce nuts, vegetables, and hybrid seeds. Sara has been working for five years with one such agrarian, Frank Muller. As president of Muller Ranch,

LLC, Frank oversees thousands of acres that grow wheat, alfalfa, tomatoes, sunflower, safflower, and almonds. Frank grew up on the farm where Sara and I meet him. His father immigrated here from Switzerland in 1947, poor but motivated. Over his life, he worked the land and was fortunate to end up with eight hundred acres that today serve as the headquarters for Muller Ranch and continue to nourish Frank and his siblings. Frank welcomes Sara and her three-month-old daughter like his own family as we climb into his pickup to learn firsthand how little things make a big difference to wildlife and farm productivity.

Pollination is a critical service nature provides to Frank's crops. Hybrid sunflowers, raised for seed that is shipped east to grow oil crops, is his largest holding this year. Bees must move the pollen from one strain to another to produce the hybrids Frank grows. But bees need more than just sunflowers, so Frank plants cover crops within his orchards and vineyards and leaves his field edges scruffy and tree-lined where possible so that bees have a diversity of flowers on which to feed. He cites the advantage of this strategy, referring to research Sara and others have conducted in the area. For example, Professor Claire Kremen from the University of California, Berkeley, worked on Frank's land to demonstrate that the cost of leaving hedgerows around his fields is repaid within seven years because of lowered expenditures for insecticides and increased seed set on crops pollinated by native bees. The bottom line is critical to Frank, but he sees beyond the dollar as well. "We just like the softer edge."

As one might expect, hedgerows that harbor beneficial bees also attract birds. A covey of plump California quail dart among the trees, confirming my suspicion as we drive to a favorite line of oaks that guards one of Frank's fields. Here and at similar field margins throughout the valley, Sara and her colleagues counted two to three times as many bird species and three to six times' higher total bird abundance than they did along fields without woody borders. Birds were especially attracted to borders that were a moderate distance from larger forests. Field borders near existing forests drew few birds, mostly because there was excellent habitat in the nearby woods. In contrast, borders far from woodlands attracted few birds because of their isolation. But hedgerows, riparian plantings, and

Hedgerows in California's Central Valley provide critical habitat for a wide diversity of birds.

widely spaced tree lines that were between just under two to three miles from woodlands attracted the highest number of birds and variety of bird species. This simple way to share the farm with native birds is especially useful in places like the Central Valley where uncultivated forests, wetlands, grasslands, and woodlands have been significantly reduced. Similar results have been obtained from studies in landscapes dominated by agriculture in Europe and the midwestern United States.

As hedgerows benefit birds, so too do birds benefit farmers. To measure the size of this benefit, Sara and her colleagues studied the influence of hedgerows on birds and invasive weevils in alfalfa grown by Frank and other farmers in the valley. Alfalfa is an important crop harvested to feed dairy cows and other livestock. It also provides important habitat for rare open country birds, such as Swainson's hawk and the long-billed curlew, as well as currently common species at risk from loss of grasslands, such as western meadowlarks and savannah sparrows. Nearly 70 percent of the U.S. alfalfa crop is grown in the Central Valley, but total acreage has declined recently because its dependency on irrigation and pesticide spraying increases its cost to farmers and the ecosystem, which is characterized by sharply falling groundwater tables and chemically contaminated waterways. The extent to which native birds increase alfalfa yields by reducing weevils would increase its profitability, lessen its chemical dependency, and improve its tenure in the valley.

Sara's team erected wire cages that excluded birds from small patches within thirty-two alfalfa fields. Comparing weevil numbers inside and outside of the exclosures provided a convincing demonstration of the birds' effects. Where birds, mostly red-winged blackbirds and savannah sparrows, could not access the alfalfa, weevil abundances were one-third higher than nearby areas accessible to birds. The stature of the hedgerow surrounding each field affected the abundance of birds and the pest-removal service they provided. Where field borders included more and taller trees and shrubs, bird attendance and the associated pest removal were greatest. But even simple field borders supporting two or more trees or shrubs averaging at least five feet in height attracted enough birds to reduce weevils from the fields appreciably.

Frank is keen to add wildlife habitat to his land wherever it makes sense. He knows the benefits scientists have shown, but in some places, it's just easier to not farm. We come around a corner, and Frank spots a pair of ducks on a small pond. He explains that this area was always swampy in the spring as winter snows melted in the high country. Farming the area, which his brother dubbed the "Mekong Delta," required annual draining. Rather than fight the muck, Frank turned it into a pond. The simple act of not farming was a win for farmer and wildlife. As Frank put it: "What used to be a struggle isn't there anymore. We really didn't give up much." Farther along the farm road, Frank spots a Cooper's hawk as it wings into a brushy peninsula jutting into a slough. This area, which Frank calls a former "point row," was also hard to farm. The triangular shape of the peninsula meant that although a long row of crops could be planted down the middle, the adjacent rows were shorter as they tapered away to untillable ground. Maneuvering a tractor down the peninsula was challenging—implements could easily be damaged when they hit hard ground bordering the crops.

Rather than risking the equipment, Frank cut off the point to create a rect-angular field and let the trouble spot revert to wild cover.

The enhancements of wildlife habitat that Frank and his siblings undertook were initially supplemented by cost-sharing grants from federal and state agencies. Now, turned off by the paperwork, they just do it on their own. And the neighbors are taking notice. One has put in a pond sur-rounded by oaks, which bustle with woodpeckers, orioles, raptors, and quail. Frank leases much of the land he farms and does what the owners stipulate, but they often ask him to treat their holdings as he treats his own. Cover crops, tree lines, and grassy verges are frequent requests. Sometimes it takes a while. One owner didn't understand why Frank let the purple needle grass and wild oats grow between his fields and the road. He in-sisted that not be done on his land. "I just don't like it," he stated, "Spray it all." After noticing the reduction in dust along Frank's roads relative to his, after two years the owner changed his mind. "I've been watching what you've done. I kinda like the way it looks. Would you do it on our land?" Frank was happy to oblige.

Frank uses an interesting combination of high and low tech to get the most out of his fields. Tomatoes are his cash cow, and high yields depend on precise irrigation. Frank and his team arrange daily watering schedules for 170 fields across several thousand acres using information on tempera-ture, humidity, and plant size. Water is delivered to each plant under the soil through drip lines, which cost about $1,000 per acre to install. To be most effective, the placement of lines has to remain precisely in the middle of each row, ten to twelve inches below the soil surface. After harvest and before the next planting, rows are disked, and the new seedbed can be centered right over the lines because their location is precisely mapped and known by the global positioning device that guides Frank's tractor. Drip irrigation has cut Frank's water use by 40 percent and increased his tomato yield by 25 per-cent, which together repays his installation costs in a single year. Add to that his savings from reductions in pesticides courtesy of the predatory lady beetles and birds his hedgerows harbor, and you can understand why he and his family have been able to remain on the land his father so proudly settled.

Frank is eager to show others how he stewards his land. National brands, such as Ragú and Sriracha, buy his tomatoes and peppers for their sauces. They love his approach and are quick to highlight his approaches in marketing campaigns. He often hosts field days at the farm to demonstrate his approach to extension agents, researchers, and other professionals. The day after Sara and I met with him, several state legislators who fashion California farm policy were stopping by for a tour. The U.S. undersecretary of agriculture will be visiting the week after to talk about the Trump administration's trade and tariff policies. Almond growers like Frank are suffering from policy uncertainties that drive nut prices down. He is confident that those who visit leave with a better understanding of the challenge farmers face and some of the simple ways farmers and nature can mutually benefit one another.

IN THE INTENSELY farmed valleys of California, farmers and scientists have taught me that the academic debate about sharing or sparing land is simply that, academic. In the places I visited, farmers share the land with wildlife because they recognize the economic, ecological, and aesthetic value in doing so. The land is spared when larger reserves are needed to protect water supplies or viewsheds. But sparing is the exception here where land values are high and nearly every workable acre has been farmed for generations. Land sharing also reflects an ethic of land use that is perhaps better tuned to the New World. When Matt Johnson visited South Africa, he found their model of land sparing to be unsettling. "Why should we put nature here and people there?" he asked. Rather than protecting biodiversity in parks and letting humans dominate everything else, he considered it better to "live with nature." And this theme kept returning to him whenever he spoke with American Indians. With these audiences, the first thing he often heard was "humans are part of nature." I heard the same sentiment as I spoke with winemakers and tomato farmers. With time, Matt has come to see land sparing as akin to apartheid. Rather than segregating people, sparing divides nature from people. I don't disagree that sparing lands from human domination is crucial for some life forms, but just as we have witnessed the inherent instability of apartheid, I worry about the stability of nature on a planet where humanity is distanced from its services.

The ability of simple actions to increase biodiversity around the margins of a landscape so thoroughly dominated by agriculture as I witnessed in the Napa and Central Valleys of California gives me hope that similar results can be obtained through land sharing actions elsewhere. Surveys that both Matt Johnson and Sara Kross have conducted suggest that many farmers are willing to pay to enhance their lands for wildlife. However, Sara is quick to note "that it really helps when wildlife provides a service." I suspect that this winning combination for farmers and wildlife will be an especially important motivator where profit margins are slim. As the studies in California have shown, however, welcoming predatory birds and pollinating insects onto the farm quickly repays the costs of sharing marginal lands, enhancing the suitability of field borders to wildlife, or erecting nesting and perching platforms. Immediate repayment comes in many forms—fewer pesticide purchases, reduced workload, lowered fuel costs, and higher soil quality. Future dividends include the pride in being able to show others a vibrant land where nature and humans can coexist. In both the short- and long-term, it seems to me that the connection between the services an ecosystem and its wildlife can provide to humans is tight when we share the land but less so when we spare it. The power of doing both can be astounding.

WESTERN WASHINGTON'S SNOQUALMIE River is restless. In this rainy country, rivers often flood and in so doing cut new channels through their rich, broad valleys. Graceful arcs of old waterways long ago, pinched off by a river's alluvial processes, dot the landscape as oxbow lakes. Small rural communities and the burgeoning suburbs of nearby Seattle have tarnished the Snoqualmie a bit and challenge the survival of its iconic salmon and steelhead. Changes in the land confront the river with pollutants, altered flows, and increased temperatures. Modern salmon runs have declined, but wild fish still navigate the river's changing course as they struggle from the Pacific Ocean to their breeding grounds within mellow reaches of the main stem and its higher-elevation tributaries. In part, the ability of rural rivers, such as the Snoqualmie, to foster runs of anadromous fish depends on the actions of small organic farms that work their floodplains. These farms improve the quality of the waters they straddle by reducing nutrient and sediment runoff through careful management of fertilizers, judicious tilling of the ground, and covering fields with winter crops. Their retention and restoration of shade trees and brush along the riverbanks cool the water's temperatures to suit salmon and trout and offer refuge to a large number of resident and migratory birds, amphibians, and mammals.

Oxbow Farm & Conservation Center is particularly proud of its land practices and the influence of these on the Snoqualmie River. Tom Alberg Jr., a successful and ecologically minded Seattle lawyer and venture capitalist, leased the 243-acre farm from his father in the late 1990s and set about restoring native plants and converting the one-time cattle ranch into an organic farm. He and his wife, Judi Beck, founded Oxbow Farm & Conservation Center in 2009 to increase the impact of the farm and understand how it might contribute to the social and natural fabric of the local area. Their vision and financial support enabled the farm to diversify and

Previous page: A male common yellowthroat perches among the grasses and vetch that serve as cover crops at Oxbow Farm.

expand its reach quickly. Oxbow produces a wide range of organic fruits and vegetables—root crops as well as leafy greens, peas, beans, and berries—that gross about $500,000 a year. But that is just the start. Oxbow also provides educational programs for local schoolchildren and in a state-of-the-art greenhouse propagates 150 species of native plants that are used for in-house and contract restoration projects. The permanent staff of twenty includes accomplished farmers and mechanics, as well as a botanist, a restoration ecologist, and an executive director, each of whom holds advanced degrees. The team works closely together to experiment with ways to sustainably spare and share land with wildlife. So, when restoration ecologist Matt Distler invited me to survey the birds on the farm in 2016, I jumped at the chance.

Oxbow includes a variety of lands that birds and other wildlife may find suitable. Thirty-four acres are actively farmed or left fallow each year, and nearly 50 acres are covered with buildings, roads, and other infrastructure. Over half of the land (160 acres) is maintained as a forest reserve or is unkempt land bordering active fields. Matt and his team have already replaced invasive, nonnative plants with native shrubs and trees on 14 acres that abut the Snoqualmie River, and they are busy restoring the remaining 30 or so. This is all part of a comprehensive environmental management plan that guides work on the farm. The plan identifies significant native vegetation on the ownership, the wildlife and ecosystem function it supports, and how best to sustain and integrate it with crop production. Because the lands at Oxbow include those shared with wildlife and spared on its behalf, assessing the birds throughout the ownership affords me a unique way to compare how these typically divergent strategies individually and together contribute to wildlife conservation.

During our twice-weekly visits to the farm from April through August 2016, my former graduate student Lauren Walker and I noticed 78 species of birds. These casual observations reveal an ornithological treasure—nearly a fifth of all 438 species ever recorded in Washington's diverse lands (ranging from grassland to temperate rainforest to tundra) and waters (both fresh and marine) can be seen in a single summer at Oxbow. Hawks, eagles, owls, and waterfowl are abundant, but we concentrated our efforts on assessing more formally the breeding assemblage of smaller birds. To do so, Lauren conducted standardized surveys within thirty-one rectangular plots of approximately six hundred square yards each, once per month during April, May, June, and July. During these 124 surveys, each of which lasted five minutes, she heard or saw a total of 40 species. These data allow us to objectively compare bird responses to land that was farmed, spared in its natural state, or shared with wildlife.

As expected, the fields at Oxbow, both actively farmed (about twenty acres) and either planted with a cover crop or left bare to fallow (about fourteen acres), held the least appeal to birds. During our formal surveys, we observed only ten species of birds in and above the fields. Over 40 percent of the individuals seen were swallows, most commonly tree swallows working the clouds of flying insects like a baleen whale works krill. The farm field community was distinct from those characterizing adjacent buffers and nearby forests. We rarely found birds typical of forest and brush in the fields, and we detected two species—brown-headed cowbird and killdeer—only in fields. Providing the open setting that killdeer, a common shorebird, requires for nesting is important, but supplementing cowbirds with the goods of the farm is a bit concerning because this native species is a nest parasite. That is, rather than building a nest of their own and raising their own offspring, female cowbirds sneak their eggs into the nests of other birds, which then raise the baby cowbirds often to the peril of their own offspring. This dynamic harmed some of the less common warblers, flycatchers, and buntings that nested at Oxbow. The single lazuli bunting nest we found, for example, was abandoned soon after a cowbird egg appeared in it.

The field buffers, comprising just over forty-three acres, including the fourteen that Matt and his crew have restored by removing invasive blackberry and replanting native shrubs and trees over the last fifteen years, represent land that Oxbow shares with wildlife. These field and river borders hosted the greatest diversity and abundance of birds we encountered. The border community included many birds that we also found in the fields but entailed twenty-seven species. Restored and unrestored borders held similar numbers of species, and both provided an important refuge for sparrows, hummingbirds, thrushes, finches, swallows, and warblers. Savannah sparrows, a widely distributed grassland species, occupied both the fallow fields, where they nested, and the borders, where they foraged and sought cover. While the species found in the habitats shared with wildlife by the farm were common and adaptable, they were distinct from those using the habitats spared altogether from farming. Nearly half of the species (thirteen) found in the borders were not found in either the upland or riparian forest reserve. These included the yellow warbler, western tanager, barn swallow, and Bewick's wren.

Two substantial quantities of the farm are spared from development, though they are explored by young students during some educational programs. The largest of these reserves spans 110 acres and supports an eighty- to one-hundred-year-old coniferous forest, typical of uplands in this region. This forest is primarily fir, hemlock, and cedar. The smaller reserve is an 8.5-acre mature cottonwood riparian forest circumscribed by an oxbow of the Snoqualmie River. Entering these reserves, one is completely isolated from the bustling farm life just a few yards distant. The dense tree canopies shade the luxuriant undergrowth of native shrubs, keeping them moist, cool, and quiet. Salmonberry, a prolific native fruit, and ferns are everywhere. Swainson's thrush, song sparrow, Pacific wren, spotted towhee, and chickadees (both black-capped and chestnut-backed) are common. Overall, we recorded twenty-three species in the reserves, the majority (fourteen) of which were also noted in the buffers. We found nine typical forest birds only in the reserves. These included eight sensitive forest-dependent species, such as brown creeper, Wilson's warbler, black-throated gray warbler, and Pacific wren, as well as the newly invasive barred owl. Each type of forest provided for a somewhat distinct avian community. In the upland conifers, Pacific wrens, chestnut-backed chickadees, and Pacific-slope flycatchers were common, while in the deciduous riparian forest we found an abundance of black-capped chickadees, rufous hummingbirds, and Wilson's warblers.

Counting birds is a quick way to assess how wildlife uses the various lands that comprise a farm, but it provides only part of the story. To gauge the farm life of birds more fully requires a hands-on approach.

THE DELICATE MIST NET is energized as a small bird senses our approach and begins to struggle. It is mid-April 2016, and Lauren and I have enlisted volunteers to help us work a trap line. We have strung a half-dozen nets, a typical tool of the field ornithologist, in the woods and brushy field borders of the farm to capture birds for our study and tag them with uniquely numbered federal and easily recognized colored plastic bands. The American robin in our net this morning inhabits all parts of the farm. Pale gray feathers on her back and russet ones on her belly tell me she is a female in at least her second year of life. Cradling her in my left hand with her neck secure between my index and second finger, I blow on her belly. The patch of red and bumpy skin I reveal is her brood patch, and it is developed enough to indicate that she is in breeding condition and most likely has eggs in her nest. The brood patch has blood vessels near its surface that impart their warmth to eggs during incubation. With special pliers, I open the small aluminum federal band and slip it over her right leg before crimping it back to form a circular bracelet. I slip a red plastic band above the metal one and fit a red over blue band on her left one. This combination allows me to identify her by getting a good look at her with my binoculars. Next week, I discover her nest in the thick blackberries that edge the river. Throughout the summer, I often watch her cock her head as she listens for worms that share the soil with crops of carrots, beets, and kale. She is a diligent mother who is paired with a male we were unable to catch and band. Together they raise at least two young to independence.

Banding birds helps us assess their performance during the nesting season on the farm. We want to make sure that those we count on a survey are not failing to reproduce or dying at an unusual rate while they carry on their lives among the farmers and machinery that work the land. In total, we banded sixty-six birds of thirteen species and followed the breeding activities of forty-eight pairs. Keeping up with the secretive family life of tiny birds that flit among thick underbrush is a challenge. However, the birds we have banded are pretty easy to spot as they sing from exposed perches or hunt the grassy field edges for insects. We use this conspicuous activity to our advantage, noting where we see particular birds and if they are in the company of a mate, carrying nest material, gathering food for offspring, or tending a free-flying family (fledglings). These clues let us piece together the reproductive success of each bird, even if we don't spy their nest.

Our brief study is far from definitive, but our results are encouraging. We confirm that at least twenty (42 percent) of the pairs successfully fledge young. Only eight of the 23 nests we observe (35 percent) failed before the young gained their independence. The nests and territories that we could reliably monitor within the cottonwood forest fared a bit better than those within the field borders. The chances of a pair fledging young in the forest were very high (82 percent of seventeen territories and nests), whereas that of birds in the buffers was slightly less (75 percent of twenty-four). While comparable, both rates are likely overestimates of the absolute rate of success. We often failed to determine if a pair was or was not successful (26 of sixty-nine instances, or 38 percent), and it is harder to confirm lack of young birds than it is to notice their occurrence.

Most of the birds we're watching on the farm are also familiar denizens of the nearby suburbs that I've spent a good part of my professional life studying. I note a few interesting differences, such as the change in common swallows and hummingbirds from suburb to farm. On the farm, perhaps in part because of the proximity to the river, tree swallows command the sky, while in the suburbs it is the violet-green swallow that dominates. Similarly, a larger body size, more varied diet, and perhaps greater tolerance of humanity enable Anna's hummingbirds to rule suburban flower patches, whereas on the farm that honor falls mostly to the migrant rufous hummingbird. On the farm, there are also some entirely new faces, and one, in particular, captures my attention.

Invisible to me, yet unmistakably from deep in the thick grass of a fallow field, pours forth a song that I can't immediately identify. The *witchity, witchity, witchity* is familiar, but not a song I know from the nearby 'burbs. Slight variations in the song are streaming at me from all corners of the farm—from the fallow land, the restored and unrestored field borders, and even the cottonwood forest. I wade into the fallow field, rank grass up to my chest, to try and glimpse the singer. As I close in, the song abruptly stops, and in its place, I start to hear a rather aggressive, throaty call note that sounds like rocks knocking together: *tschik, tschik, tschika, tschika.* Finally, I see the little guy that has so much to say. He is gripping two long blades, one with each foot, doing the splits, bobbing and weaving as a slight breeze

moves the grass. His black mask, yellow breast, and white forehead are un-
mistakable. I usually associate the common yellowthroat I'm ogling with
cattail marshes, yet here on the farm, they seem quite at home wherever
thick vegetation inhibits my movement. Indeed, as the summer progressed
Lauren and I observed yellowthroats in all the habitats of the farm: forest,
fallow field, and scrubby borders. Only the American robin was equally
ubiquitous.

The Pacific coast race of the common yellowthroat, which inhabit the
farm, winters in southern California, northern Mexico, and Arizona. These
small warblers (the six we caught ranged from a quarter to just over a third
of an ounce, somewhat less than a typical house key) migrate hundreds to
thousands of miles to their breeding grounds, which stretch from northern
California to southeast Alaska. Males are said to arrive for breeding about
a week before females. By the time we started netting in mid-April, both
males and females were on the farm. All the males we caught were in breeding condition, as indicated by complete cloacal protuberances. The clo-
aca of a bird ends in the vent, a multipurpose opening from which wastes,
sperm, and eggs exit the animal. As males ready for reproduction, the area
just upstream of the vent swells into a BB-pellet-sized bulge, which we call
the cloacal protuberance. Swellings are less pronounced on females. However, their readiness as a parent is indicated by a naked and vascularized
belly. Neither female we captured had yet developed this brood patch required to nurture the eggs. Persistent singing suggested that the males were
busy advertising their wares and claiming a spot of ground, while females,
who likely just recently arrived, refueled after their migration.

The chorus of yellowthroats waned as April gave way to May. Males,
having parsed the farm into abutting territories and attracted mates, spent
each day finding food for themselves and their females. The females vanished from our observations for about ten days, during which time they
occupied the nest they had built and sat on the eggs they had laid, quietly
incubating. I looked intently in the overgrown vegetation, where I was sure
a nest must be, but was unable to find any. Rather than risk trampling the
nest with persistent searching, I was glad to monitor their nesting activities
by merely observing the parents' behaviors. Watching both members of a

mated pair stuff their beaks with insects and then disappear into a thicket told me that a nest was not only nearby but also harboring hungry chicks. I'd learn more if I could count the chicks or measure the eggs, but just knowing that the yellowthroats were making a living on the farm was what really mattered to our study. Besides, I'd get my count after the young left (fledged) the nest.

Throughout June the yellowthroat families remained within thirty yards of the presumed nest site. I focused my attention on two families that sheltered within the thick shrubbery buffering the working and fallow fields from the river. Lauren watched four other families, all of which lived entirely within the larger riparian forest reserved by the farm. Throughout the month we occasionally glimpsed the fledglings as their begging was rewarded with mouthfuls of insects delivered by their mother or father. The families remained ensconced within the thick cover, relying on the habitat spared from cultivation. Our conservative estimate was that three pairs succeeded at fledging young; those for which we had no evidence of successful reproduction lived within the riverine forest, where they typically skulked low to the ground, well hidden from our curious eyes. Some of these sneakers were likely successful. Regardless, our birds seemed to breed as well as those elsewhere in the species range (other studies report that a third to a half of nests fledged young).

In early July the two families I watched moved out of their natal thickets and into the tall grass and vetch of an adjacent fallowed field. There, I could more easily observe the fledglings, which were plumed similarly to their mother and lacked the strong black mask of their father. By resembling females, the increasingly independent offspring were less likely to draw the ire of neighboring males who loosely defended their territory boundaries. The fallow field offered the shelter and food that a yellowthroat family required. Simply not mowing, grazing, or tilling the ground enabled grass to grow over four feet high and provide excellent cover for birds, especially young yellowthroats whose dull yellow, green, and gray plumage blended perfectly with the senescing grass. Eschewing pesticides and planting flowering cover crops to sustain beneficial insects also played right into the yellowthroats' hand. Throughout July and early August parents and young alike feasted on the insect bounty within the grass, fattening up for the autumn migration that would take them back down the coast to their wintering grounds in California, northern Mexico, or possibly Arizona.

Adam McCurdy, the lead farmer at Oxbow during our bird study, was happy to let the grass stand throughout the summer for the yellowthroats. To Adam, a "fallow field is a canvas" where he works to build organic matter and nutrients. But Adam typically mows the fallow fields to keep invasive blackberries in check. Not mowing increased the risk that these and other weeds would establish, spread, and prove challenging to control in subsequent years. Mowing also helps limit crop-hungry insects, such as flea beetles, cucumber beetles, diamondback moths, aphids, cabbage moths, and cabbage loopers. These pests damage Adam's vegetable crops by chewing leaves and slurping critical plant fluids. I figured the yellowthroats would help by eating many of these pests, but Adam took no chances. He sprayed his kale, cabbage, brussels sprouts, and other brassica vegetables with the bacterium *Bacillus thuringiensis* (Bt), which produces proteins that are toxic to the larval moths and loopers. Adam's second defense against pests is not typical. He and his workers remove beetles from their crops with a vacuum. A leaf blower, run in reverse, makes a lightweight and powerful suction device. No doubt the machine's manufacturers did not anticipate this application of their tool, but it draws up beetles like a champ and readies the fields for the next weapon, the release of beneficial insects. By mid-July, Adam is hatching his third batch of green lacewings and parasitic wasps that destroy larval loopers and moths. In addition to the vetch within the grass, field borders of sweet alyssum and psyllium provide habitat for the beneficial insects. So, while the yellowthroats work the grass, the farmers work the crops to limit the spread of harmful insects that escaped the maws of birds.

Adam's crew took care of the blackberry by mowing the tall grass after the yellowthroats migrated. Throughout the winter the clipped, but still fallow, field attracted a variety of wild animals. By midwinter, the grass was crisscrossed with runways indicative of voles that eat and wear down the turf just below the winter's snow. Among these well-worn paths, which remind me of little bobsled tracks, were signs of predators. Coyote scat and

Farmer Adam McCurdy and his son enjoy working manure and compost into the fields before planting.

footprints told the story of one hunter that worked the grass for voles and other rodents. I suspect the short-tailed weasel I often flushed from the shrubby field borders also stalked voles. Indeed, the resident red-tailed hawk and northern harrier benefitted from the voles' abundance.

The fallow field is not the only green on the farm during winter. In fact, there is no bare earth visible except for the few two-track roads used to access the fields. Each active field is protected from the typical heavy winter rains with a short growth of vetch and clover that also provides flowers for pollinators and root nodules harboring particular bacteria that convert atmospheric nitrogen to other forms, including ammonia, that will fertilize next season's vegetables.

The green of winter is a valuable resource for wildlife well beyond the vole ecosystem. In February, I am treated to a symphony of goose music. Small cackling geese, hundreds of them by my count, wheel above the farm honking in rapid soprano voices. Their larger cousins, our familiar Canada geese, stream by in loose V-formation and add bass to the tune.

Both species frequently graze on Adam's cover crops, adding their drop-
pings to the nutrient stockpile. Trumpeter swans and snow geese also see
the cover crops as a welcome buffet. All are tolerated and enjoyed by the
farmers at Oxbow. Two other winter residents, fox sparrows and golden-
crowned sparrows, also venture into the fields from the adjacent shrubs.
Through early spring they mingle with the year-round resident song sparrows and white-crowned sparrows.

The progression of foods juxtaposed with the cover that Oxbow provides makes it a nexus for migration. Species able to eat plant materials during the winter, including the sparrows, geese, and swans that breed in the
Arctic, flock to the farm in autumn, just as the migrant swallows, thrushes,
and warblers that rely mostly on insects jet off to their southern wintering grounds. In spring, as longer days thaw the north and spur on insect
blooms on the farm, the southern migrants return to breed just as the birds
of the tundra head north. The farm is especially attractive to migrants that
arrive in the spring to breed. Of the seventy-eight species we encountered,
twenty-eight reside at Oxbow only during the spring and summer. Only
seven species spend winter on the farm before flying north to breed. This
annual parade of bird species—supported by the cover crops, native plants,
and associated invertebrates they sustain—is one reason that the farm can
host so many species.

THE IMPORTANCE OF farms as wintering habitat for wildlife is widely recognized. In Britain, for example, a dozen species of waterfowl, two species of
grouse, and woodpigeons graze on grass, young cereals, and crop residues
during the winter. They are joined by more than two dozen other birds, including plovers, snipe, gulls, owls, woodpeckers, and many songbirds, that
feed on soil and surface invertebrates. In addition, weed seed and waste
grain are gleaned by at least sixteen species of partridge, dove, lark, sparrow, finch, and bunting. As farming intensified from the 1930s onward, the
stubble, fodder roots, spring tillage, hay piles, and stockyards that provided
most of the resources required by the birds declined. Not surprisingly, the
diversity of birds wintering on British farms has also decreased.

The wheat and pasture grass farms east of the Cascade Mountains in

Families of trumpeter swans are common winter residents on farmlands where cover crops are planted and some grain is left unharvested in western Washington.

Washington always surprise me with their abundance of wintering raptors. Each March, I drive with students from George to Spokane, a distance of 132 miles, on the way to a week in Yellowstone National Park. We stay alert through the seemingly monotonous agricultural landscape by counting hawks and falcons. We routinely spot two species of sit-and-wait hunters (red-tailed and rough-legged hawk), four pursuers (merlin, kestrel, northern harrier, and Cooper's hawk), and a couple of chaser/scavengers (golden and bald eagle). Our count in 2018 was typical: seventeen red-tails, six kestrels, four rough-legs, three harriers, a bald eagle, and six sit-and-wait hawks we could not identify to species (birding at seventy miles per hour is not an exact science). All of these apex predators gather on the wide-open land to feast on rodents, rabbits, and smaller birds that are abundant in the stubbled and crop-covered fields, as well as the occasional road-killed deer that misjudges its crossing of the busy highway.

In California, some fields are uniquely prepared for wintering wildlife. Rice, despite intensive cultivation during much of the year, is left to the birds during winter. Fields of clay are precisely groomed and leveled by farmers each spring so that they hold water at a constant level of five inches during the planting and growing season. As the plants grow and grains mature into the autumn, the fields are drained for harvest in September. In the past, farmers burned the rice stubble, but concerns for deteriorating air quality led the state to ban most burning. Now many farmers flood their harvested fields to aid the decomposition, which enables waterfowl and other birds, mammals, reptiles, and amphibians to use the grounds throughout the winter. As rice straw breaks down, it supports a sumptuous buffet of bugs, which grebes, pelicans, herons, ducks, geese, hawks, rails, shorebirds, gulls, owls, doves, swifts, kingfishers, flycatchers, swallows, corvids, and many songbirds flock to for feeding. The bevy of birds is joined by over a dozen mammals, including bats, rodents, coyotes, foxes, skunks, otters, deer, and beaver. Frogs, toads, lizards, turtles, and snakes are also common. In all, nearly 230 species of vertebrates have been recorded using California rice fields.

Biologist Rodd Kelsey and restoration ecologist Sara Sweet introduced me to some of these fields in the Sacramento River Valley. Here, in the northern part of California's Central Valley, a mix of private and federal organizations, notably, the Nature Conservancy (which employs Rodd and Sara), Ducks Unlimited, Audubon California, and the U.S. Bureau of Land Management, cooperate to provide thousands of acres of wetlands for wildlife. Some of the focus is on restoring valley bottomland forests and, in other areas, on giving wetlands deeper water needed by wading birds and some ducks, but increasingly managers are also focused on providing the rich, shallow waters that are necessary for migrating shorebirds and cranes. About 250,000 acres in the northern valley are spared from agriculture as wildlife reserves. But the remaining 3.75 million acres formerly available to birds are now farmed. Purchasing the additional land that wildlife needs is prohibitively expensive (providing critical links alone is estimated to cost $3–$5 billion), so the Nature Conservancy and its partners are working to learn best how to share agricultural lands with wildlife.

As we climb the dikes that mostly keep the peaty soils of the conservancy's 9,200-acre Staten Island from being inundated by the Mokelumne

River, we look over vast fields of wheat and corn. Although these crops are farmed conventionally for profit, some practices are being tweaked to support the island's large wintering population of greater and lesser sandhill cranes. For example, after the corn harvest, the residual stubble is chopped and rolled before being gradually flooded. Chopping and rolling increases the availability of waste grain, and the measured, sequential flooding provides an ever-changing, shallow shoreline that is the favorite feeding ground of the cranes and shorebirds. Sharing with wildlife doesn't just occur on Nature Conservancy lands. By partnering with Audubon California and Point Blue Conservation Science (a not-for-profit conservation science organization), wildlife-friendly farm practices such as these are being exported throughout the valley. As Rodd describes, conservation biologists are unusually doing this, by creating an "Airbnb" that rents farmland to birds.

Renting farmland begins with an auction, done in reverse. Farmers pitch bids to a single buyer, the Nature Conservancy, indicating what amount of money they would accept in payment for leaving water on their rice through the winter. Delaying the drying of rice fields can be risky; if spring rains are heavy, fields may not be sufficiently dry for the plowing that precedes summer flooding and planting. Rodd explains that determining which bids to accept is a precise, high-tech undertaking. The conservancy does not merely take the lowest bids but instead seeks to identify and rent lands where the birds are and where the wetlands are not. Biologists consult weekly models of bird occurrence derived from Cornell University's e-Bird program. e-Bird takes records of where birders find birds and amasses them into a giant, time- and location-specific database. Scientists then use the citizen-driven database to model bird occurrence throughout the Sacramento Valley. Next, the locations of bird occurrence are combined with maps of where flooded fields are most available (or not) developed in collaboration with Point Blue using satellite images of the valley. The team puts both maps—birds and wetlands—together to determine where, during any two-week interval, there exist "bottlenecks between what habitat is provided by agriculture and what birds actually need." The Nature Conservancy accepts bids from farmers in those bottlenecks to annually provide ten thousand acres of relevant and high-quality shallow wetlands to the ibises, plovers, stilts, and sandpipers that winter in a land that each year also provides nearly five billion pounds of rice.

The eastern sky casts a purple glow over the Sierra Nevada as a golden sun slips below the Coast Range to the west. The flooded rice, alongside which I will camp tonight, turns golden as flocks of greater white-fronted geese gab and cartwheel from the sky into a nearby field. Marsh wrens, black-necked stilts, and killdeer rally as the Sacramento Valley darkens. In my tent, as the dry air quickly turns cold, I watch the Big Dipper rotate through a clear sky and review the many ways farms in this valley share their goods with birds. Flooded fields provide food and safety. Hedgerows provide cover for a wealth of sparrows and enable birds from the savanna, such as the Swainson's hawk, to find suitable nest sites. Grasslands, though mostly composed of species from the Mediterranean, such as wild barley, harbor unique vernal pools that, as we'll see in chapter 10, can benefit from grazing. Throughout the night I am awakened by the hoots of a great-horned owl and the distant howls of coyotes. The din of Pacific chorus frogs lessens as the sky lightens ever so slightly at 5:00 a.m. The geese kick off the predawn symphony, but they are soon joined by the quacks of mallards, whistles of cinnamon teal, and sharp clinks of American avocets and black-legged stilts. I rouse from the tent, slip on my coat and stocking cap, and bird the borders of a series of rice fields.

Ring-necked pheasants announce their presence with roosterlike crowing as I move away from the tent. A huge roost of tree swallows begins to stir, feasting no doubt on the many aerial insects that swarm through the air. A common raven wings over me, uttering *quorks,* a call I know the species uses to defend territory. In the half hour before sunrise, I log twenty species. I edge up to a distant field and watch three stilts glide through the shallows like figure skaters on long, pink legs. They are soon joined by a small cluster of white-faced ibis. Those in front of me are nonbreeders, sporting drab brown feathers flecked with white on their necks and heads and showing just a hint of green sheen on their wings. Soon there are a dozen ibis, as the flock attracts several bronze-bodied adults with iridescent green wings, red eyes and legs, and reptilian bare, reddish faces outlined

White-faced ibises and black-necked stilts forage in a flooded rice field.

in white. The whole group of ancients probes their long, curved beaks into the water's mucky bottom. That I'm in the company of ibis is a testimony to how the valley has changed. Ibis numbers crashed in the 1970s because few wetlands were maintained, but now their population is booming because of agricultural wetlands. This vibrant scene is far from anything I imagined could exist in a landscape dominated by expansive farms, vineyards, and ranches. In my short twenty-four hours, I tally seventy species of birds, including a male common yellowthroat that sings from a small patch of cattails bordering the flooded rice.

AN EARLY SPRING SUN cuts through thick river fog to brighten up Oxbow on the last day of March 2017. I'm excited to see if the yellowthroats that ruled the grass last autumn have returned. Migrant swallows cut arcs through the sky, but the cold day offers them few insects. Robins are everywhere, singing and sporting their spring finery that includes a bright yellow beak, a white eye ring and vent, a black cap and tail, and of course a rusty red breast. I spy female robin red/blue, right back on her territory from last season. A banded song sparrow that lives next door has also survived another year. A belted kingfisher rattles overhead as it cuts across the farm between bends in the Snoqualmie River. Not to be outdone, a male Anna's hummingbird rockets up to a height of 150 feet, then plummets toward the ground with such force that the air blasted through his tail produces a high-pitched, screeching *POP*. His flight path traces a narrow "U," whose curved base points to a honeysuckle thicket where I suspect a female may be nesting. Despite all the avian excitement, I don't hear a single yellowthroat.

Sarah, one of Adam's assistants, is planting and nurturing young kale, collards, Napa cabbage, chard, fava beans, peas, radishes, and carrots when I return to the farm in mid-April. In my absence, Adam has lightly plowed and disked the active fields to break up and turn under the winter cover crops, which will help fertilize Sarah's seedlings. He has also shaped each field into a series of furrows and four-inch-tall, two-foot-wide, fifty-yard-long seedbeds. The flat-topped seedbeds are light gray in color and packed just enough to seal young roots from excessive airflow that stifles their growth. The well-organized field eases the farmers' workload during planting and weeding. An old orange Allis-Chalmers tractor straddles each mesalike seedbed, and when equipped with a perfecta implement, it can weed, texturize, and reform each bed with a single pass. The perfecta doesn't deeply plow the ground but does lightly stir and break the clods from the top few inches of soil, which encourages dormant weed seeds to sprout. With each flush of weeds, Adam reshapes the bed, shaving off the top inch of soil and associated weeds. This reshaping is done three or four times before crops are planted, after which weeding depends on labor-intensive hoeing by hand. As I survey Adam's handiwork, my attention is captured by a crow diving at a red-tailed hawk. And then, I hear a familiar tune coming from the west. The yellowthroats have returned.

I hurry to the riverbank tangle to get a look at the singing yellow-throat. He is moving among the fresh green blackberry leaves and pink salmonberry blooms in a section of the hedge where one of my pairs raised a family last year. I struggle to get a clear look to determine if this male is banded, a sign that he is last year's territory holder. Finally, I see a silver glint from his right leg—at least one male is back from his winter travels

and again defending his Oxbow territory. As the summer progresses, I discover that four males have divvied up the shrubby borders adjacent to the
still fallow field, which by July again hosts a great crop of grass. I confirm
that these males accompany their families into the grass, just as they did last
year. I often see the banded and at least one unbanded male but never confirm the return of the second male or either female that I banded last year.
Either these breeders have moved to a new location, or they did not survive
the winter. The latter would be surprising, given that a yellowthroat's maximum lifespan is more than eleven years. I'll need to band more birds and
monitor for more extended periods of time to know for sure.

The Oxbow yellowthroats used the lowland reserve that was spared
from farming and the fallow fields, brushy river borders, and untidy field
edges that were shared with crop production. The ability of this migrant
warbler to thrive in unplanted snippets around a farm suggests that it might
do well elsewhere in its range where farmers share their land with wildlife. Indeed, in North Dakota, the yellowthroat occurs at modest densities
on farmland enrolled in the Conservation Reserve Program where, as in
Nebraska, farmers are paid to conserve soil and water on mainly erodible
land. Conservation strategies include planting fields and field borders with
grasses, such as alfalfa, brome, and wheatgrass, and legumes, such as vetch,
instead of annual crops. Other sensitive grassland species also benefit from
Conservation Reserve Program set-asides, including several that are declining severely across the northern plains, such as lark bunting, grasshopper
sparrow, clay-colored sparrow, and bobolink. Lands are typically enrolled
in the Conservation Reserve Program for a ten-year-long contract period,
after which farmers may return their property to crop production. The authors of the North Dakota study feared that many farmland birds would
suffer if Conservation Reserve Program lands were restored to production
at the end of their contracts. Statewide populations of yellowthroats, for
example, were projected to drop by just over 9 percent if farmers no longer
employed the wildlife-friendly approaches stipulated by the program.

THE ACTIONS OF Adam McCurdy and his staff at Oxbow harmonize with
the needs of the many birds that spend most or all of their year on the

farm. Crop rotations that include seasonal or annual fallowing with cover crops build soil organic matter, reduce the need for fertilizers and herbicides, and provide important habitat for creatures, such as the yellowthroats, waterfowl, and grassland sparrows that share the farm. This old and straightforward practice of resting the soil and rotating a diversity of crops grown on a single field is the foundation of what is called conservation agriculture. Earth scientist David Montgomery discusses at length how these practices help farmers around the world increase their bottom lines by simultaneously reducing the need for expensive chemical inputs and increasing crop productivity in his book *Growing a Revolution*. Healthy soil also helps reduce greenhouse gas emissions from agriculture by increasing the soil's ability to store carbon and, when tilling the ground is reduced, lessening the demands for tractor fuel. These natural and fiscal benefits of soil stewardship are energizing a movement among conservation farmers to have their crops, whether grown organically or by conventional methods, certified as Regenerative.

The crops grown at Oxbow are certified as Organic, and they would likely also be eligible for the label of Regenerative. Most proposals for the brand, such as one introduced into the 2016 Vermont legislature, require that soil organic matter increase over time. Not disturbing the soil after harvest is a key to building the organic component of soil because cover crops stabilize and mulch soil, which enables microorganisms to flourish throughout the year. These microscopic bacteria, fungi, and animals increase soil fertility, and by decomposing mulch, they build the soil's organic matter. Adam disrupted part of this process when he employed moderate, shallow tilling to eliminate early flushes of weeds and give his carrots, beets, and onions a competitive advantage, which they need because they have small seeds and cast little shade as they grow. He further affected soil processes when he shaped the bed and furrows to increase drainage, which is vital in the wet climate of the Pacific coast because it enables earlier spring planting. These small setbacks to soil microbes are offset by frequent fallowing and generous application of compost and manure. These amendments have helped increase soil organic matter to a level of 3.4–5.1 percent, a value similar to that attained after more than a decade of organic, no-till farming at the Rodale Institute's experimental farm.

Avoiding herbicides and pesticides clearly distinguishes the day-to-day farm chores at Oxbow from those I encountered on the intensive corn and soybean farms of Nebraska. Rather than growing plants genetically engineered to be immune to glyphosate herbicides and then spraying fields with that chemical to kill susceptible weeds, the farmers at Oxbow weeded active and fallow fields by hand or where possible by machine. (Glyphosate is the active ingredient in the weed killer known as Roundup.) Instead of planting seeds coated with neonicotinoid chemicals—known as neonics—to infuse growing crops with onboard insecticides, the Oxbow agrarians vacuumed pests from their crops, reared and released insects that attacked the pests, and provided forage plants to sustain populations of the beneficial insects. Working without chemicals is labor intensive, but the investment pencils out because increased soil health boosts farm output while lower seed, fertilizer, fuel, and pesticide expenditures reduce farm costs. It also builds human capital by employing many field hands and providing them with safe working conditions.

Refraining from the use of neonics is especially crucial to the health of the native bees that pollinate many of Oxbow's crops, including raspberries, apples, blueberries, peppers, squash, and peas. Global bee populations are on the decline, and a large body of research points to neonics as one of the culprits. As neonics disperse throughout plants, they enter the nectar and pollen that bees feed on. Ingestion of these poisons harms bees' nervous systems and has been linked to reduced growth of bumblebee colonies in Sweden and mason bee colonies elsewhere in Europe. Honeybees may also decline as exposure to neonics increases, but the evidence is less clear for this species, especially when considering experiments done in realistic agricultural settings. However, honeybee colonies seem especially prone to collapse when exposed to neonics in combination with other regularly deployed insecticides and fungicides. By avoiding neonics and other pesticides and by reducing the size of active fields and ensuring a bounty of native and planted spring flowers, the farmers at Oxbow realize a basic service from nature—pollination of their crops.

Converting field borders from blackberry brambles to diverse communities of native plants increases the resilience of Oxbow's pollinators to the vagaries of the Pacific Northwest's weather. The proximity of warm water in the Pacific Ocean to Washington's coast varies significantly from year to year, and this affects the spring and summer temperature and rainfall inland. The influence of this variation on pollinators is buffered by the many plants at Oxbow; some, such as red flowering currant and salmonberry, bloom early in the spring when queens from our many native bumblebee species are searching for new nest sites. Finding a nest site and the resources needed to fuel colony growth at this time of the year is crucial for the persistence of these and other native pollinators. Other plants, from cow parsnips on wet ground to lupines and sunflowers on drier parcels, bloom as summer peaks and fades to autumn. Having a diverse portfolio of flowering plants, rather than merely the late-summer-blooming blackberry increases the resilience and function of native pollinator communities. Supplying the plants for this endeavor takes me to an entirely different part of the farm, one mostly under glass.

BRIDGET MCNASSAR WELCOMES me to the new, 6,400-square-foot, state-of-the-art greenhouse and its row upon row of native flowers, shrubs, ferns, and trees. Now forty-two, Bridget is not your traditional farmer. She initially put her master's degree in education to work as a middle school teacher before earning another master's degree in natural resources. It is the second degree, during which she developed techniques to grow native conifer trees in a nursery, that has brought her to Oxbow. Still, Bridget is a scientist, and here in the greenhouse and adjacent outdoor garden, she studies how to sow and germinate seeds to produce vigorous native plants. She captures the natural variation within a plant species by starting with seed rather than propagating from a bit of root or stem, as is common in the nursery plant trade. Propagating from roots or cuttings simply replicates the parental stock, but propagation from seed preserves the genetic variability that may be a key to a plant's ability to tolerate new and changing conditions on the farm.

The Oxbow native plant nursery houses 150 species of plants. For many, there is no information available on how to germinate their seeds. So, Bridget sets up trials in which she treats or sows seeds in a variety of conditions. For example, fresh seed of the red columbine, an iconic wildflower of the region that works well in landscaping and attracts pollinators, has notoriously low germination rates; typically, less than half of the sown seeds grow. Bridget started to save extra seeds over the past three years after hearing from other growers that seed left to age for a year or more germinates better. She stored the leftovers and then planted the three lots—aged over one, two, or three winters—in a simple experiment. Despite originating from the same source and being handled in the same way, the seeds stored longest had the highest germination success. The nursery's columbine protocols have been adapted to account for this new knowledge. Pacific dogwood also presented problems. The species is difficult to grow in a nursery setting, so Bridget has been investigating the influence of drying versus storing seeds in a moist, cold environment before germination. Her investigations are ongoing, but the initial results suggest that drying (which is the typical way the seeds are shipped to nurseries) and lack of a sufficient cold treatment reduces germination. Bridget is constantly modifying her propagation protocols based on her experience, and she shares her best practices with others via a national native plant database. In this way, she is expanding the farm's ability to restore its lands as well as helping others across the country learn from her experiences with native plants.

Bridget's curiosity doesn't stop with a seed's successful sprouting. The goal of the nursery is to produce healthy plants for sale or restoration of lands on the farm. Although we often focus attention on the above-ground portion of the plant, what goes on below the soil is extremely important. To develop a healthy root system, Bridget has learned to start her seedlings in conical, open-ended pots. These cones promote root health by limiting primary root development, increasing desired fine roots and reducing curling typical of plants grown in flat-bottomed containers. When I last visited the greenhouse, native sedges—grasslike plants with triangular, sharply edged stems that typically inhabit wet meadows—spouted from hundreds of cones. The white plastic cones topped by wispy green blades looked like mutant carrots, but in truth, they were the start of a restoration project that involved Bridget and Matt Distler, the farm's restoration ecologist. The understated plants were soon to become the front line of a battle between Matt and one of the farm's most pernicious weeds, reed canarygrass.

Reed canarygrass, native to Europe and introduced to North America as an ornamental, forms dense monocultures in wet areas. In Washington, as elsewhere in the United States, it is classified as a noxious weed. To restore stream- and lakeside vegetation to its native state means that reed canarygrass has to go. Traditional methods of canarygrass removal include mowing and spraying with herbicides, but neither is practical or desirable at Oxbow. Pounding willow stakes — three-foot-long sections of young willow stems, sharpened like punji sticks — into canarygrass in the hopes that they will grow, develop a canopy, and shade out the grass is typical on organic farms, such as Oxbow. But Matt is trying to improve on this approach.

When he first started working on the farm, Matt noticed that in low-lying areas where the water level was a foot or two deep throughout the winter and moist all summer, two native sedges (western inflated sedge and slough sedge) seemed to be good competitors with reed canarygrass. He wondered if planting these sedges, in addition to pounding willow stakes into grass areas, would help limit the turf and add some complexity to the ground beneath the shade of the willows. And so with Bridget's ability to collect and germinate sedge seeds, Matt is testing out his idea. As early farmers split their fields into various test plots, Matt has partitioned some of the reed canarygrass areas into ten-by-ten-foot sections. Each plot is randomly selected to receive various types of weed control (woven plastic weed mat, biodegradable and flood-resistant sheet mulch, or neither), willow planting (twenty-five stakes that include either a mix of two native willow species or no willows), and sedge planting (a blend of western inflated and slough sedge, a native bulrush, and a native, common rush, or no planting). Over the next several years Matt and Bridget will monitor the growth of reed canarygrass in each plot to determine whether sedges and various types of weed control enhance the ability of willows to shade out canarygrass and return the wet meadows of Oxbow to the biological diversity they enjoyed before canarygrass moved in. That is if they can perfect a beaver-proof fence that will keep the big rodents out of the willow stakes, which are among their favorite foods. So far, the beavers are winning 2–0.

The experimental mindset of the botanists and ecologists at Oxbow has an expansive reach. Matt and Bridget regularly teach workshops, deliver lectures, and lead tours so that others can learn about their successes and failures. And Bridget's produce that leaves the farm often takes with it a native story. In 2017 the staff produced about 65,000 plants, 30,000 of which went from the farm to support local restoration work, landscapers, retail nurseries, and local conservation district needs. A crop of 14,000 native western camas lilies started in 2017 plus another 25,000 plants of twenty species now thrive at the University of Washington's new natural and cultural history museum, which opened in 2019. Here not only do they provide a touch of purple to the early summer and feed native pollinators, but they help educate students and visitors about the role camas plants had in the economy of indigenous peoples. In Oregon and Washington, the bulbs of camas were dug with canes crafted of local wood and antlers on traditional grounds by tribes, who valued the roasted bulbs for their sweet flavor. The prepared bulbs were traded throughout the region and offered as gifts at weddings and funerals. Members of the Lewis and Clark Expedition gratefully accepted camas and dried salmon from the Nez Perce tribe while camped on Idaho's Weippe Prairie in late September 1805. Though satisfying, the new diet did not sit well with the ravenous men. Many, including Captain Lewis, were incapacitated for days with extreme diarrhea. On his recovery, Lewis asserted that camas "is palateable but disagrees with me in every shape I have ever used it." Later ornithological explorers John Townsend and Elliott Coues agreed. It is too bad that these early natural historians did not live to see the good things Oxbow is doing with their most disagreeable plant!

TEACHING WITH NATIVE plants is only part of the educational mission at Oxbow.

Three groups of a dozen children are scattered about the Kids' Farm and Living Playground learning about Oxbow's crops. They giggle and jump around as they pick plant names from a hat. Scarlet runner bean. Clover. Chickpea. They learn the name, have fun saying it out loud, and then become the crop, learning and taking on the essential qualities—how it grows, what it looks like, how it tastes. Their wonder and curiosity about the cultivated world are piqued. They learn where food comes from and why farms are important. And they have a blast in doing so.

Oxbow reaches over seven thousand kids a year as their farmers speak in local schools and the farm hosts summer day camps. These activities closely link formal classroom teachers and less formal outdoor-based educators to explore plant families, decomposition, and pollination. Learning focuses on the power of observation and connecting what is done on the farm to the food that kids eat. The curriculum developed at the farm supports the Next Generation Science Standards, which emphasize K-12 learning experiences that cut across disciplines, expose students to how scientists work, and teach core concepts in the physical sciences, life sciences, earth and space sciences, and engineering, technology, and the application of science. At Oxbow, students can work with scientists to discover how farm practices affect food quality and ecosystem function. They can explore how critical observations and experiments inform plant propagation and restoration ecology. They are surrounded by the application of science to the land so that nature and humanity are well served. And they get their hands dirty.

Older students also have a lot to learn at the farm. Partnering with the U.S. Department of Agriculture, Oxbow farmers are reaching out to their neighbors to offer workshops that prepare them to be certified to sell their produce to schools and other institutions. In this way, a community of farmers can improve the ability of the foodshed—the area where food is both raised and consumed—to meet the dietary needs of its inhabitants. Throughout the United States, the ability of local farms to meet our food requirements has steadily decreased, mostly because of population growth, suburban sprawl, and loss of farmland. For example, while nearly all of the population could be fed with food grown within a fifty-mile radius from 1850 to 1900, even if all local farmland produced food for nearby people, they could only feed 80–85 percent of the population in 2000. If we expand the foodshed to a one-hundred-mile radius, most cities, including Seattle, can be supplied from their agricultural lands. The ability of local areas to nourish their inhabitants depends less on dietary choice (vegetarian versus a diet that includes meat), food waste, and crop yield than it does on social and economic considerations. To eat local, residents require a diversity of locally farmed products available year-round. This potential bounty can only be produced by farmers that benefit most from local markets rather than global markets that are today's norm.

Globalization of agricultural trade, which has disrupted traditional social arrangements and connection to the land, is recognized by many scholars as threatening to ecological and human well-being. For example, increasing global demand for shrimp has produced jobs for Thai farmers who developed rearing ponds in tropical seaside locations. However, mangrove forests—which are critical to the integrity of shorelines—were often cleared to make room for shrimp farms, and the saline water in the farm ponds reduced the fertility of nearby lands to sustain vegetables, fruits, and other agricultural products. And, according to Thai researcher Louis Lebel and his colleagues, because "shrimp farmers are the small fry in the big pond of international shrimp trade and politics," the local environment and farmer livelihoods were affected by policies and tastes in Europe and the United States more than they were responsive to local conditions. Food safety, exploitation of vulnerable farmers, and degradation of overworked farmland all occur as the distance between farm and table spans nations and continents. As Abigail Cooke and her coauthors noted, the increasing complexity and interdependence of producers, consumers, and an array of anonymous brokers, agents, and service providers increase the "vulnerability of the system as a whole . . . as recent scandals over food safety and public health clearly illustrate." By producing foods for the local market and helping their neighbors serve nearby traditional and emerging markets, such as the school lunchroom, Oxbow is helping to reduce the vulnerability of its land and community by shrinking the foodshed.

The staff at Oxbow seems always to be looking at ways to increase their connection to the community. Each weekend in October as the harvest winds down, for example, the farm throws a country bash. As my family and I enjoy local foods at the Oxtober free public event, Baby Gramps, a Seattle musician, pumps out jazz, blues, and folk music. Kids are making colorful flower headbands and carving jack-o'-lanterns. Hayrides, farm tours, a huge pumpkin slingshot, and a playhouse made of hay entertain and inform adults and kids alike. Old hippies and young urbanites pick tomatoes and pumpkins or buy already harvested veggies at the farm stand. And the birds pay us little mind. Robins and blackbirds are taking grapes and the few remaining raspberries. Juncos work the furrowed field. Squadrons of waterfowl cruise high above the river. On this crisp autumn day, the farm offers more than food and fun. It demonstrates how land that values nature can also nurture people.

THE SHARING AND SPARING of land practiced at Oxbow weave a rich tapestry of grassland, cropland, brush, and forest, which are used by an impressive diversity of resident and migratory birds. Both strategies work well together to provide areas for somewhat sensitive species as well as those that better tolerate human activity. I have learned to see these two approaches to conservation as complementary rather than exclusionary. Often which tactic is employed merely is a matter of land availability. If space and resources permit farmers to leave some of their lands in a natural state while they also apply wildlife-friendly practices in their fields and pastures, then native fauna from birds to insects and fish to mammals will find a place to thrive.

The choice to share the farmland and spare some wildland on this relatively small farm was a decision made by the owners that reflects their principles rather than a response to incentive programs or regulations. Oxbow Farm & Conservation Center was founded as a nonprofit organization that has a diverse mission of conservation, education, and food production. The farm values stewardship of nature equally with maximizing the harvest. Yet despite a range of activities that might be seen as contrary to the goals of a typical farm, each acre in production grosses around $20,000 per year, which just covers the cost of farming. Though not turning a profit, the farm is increasing the diversity of sustainably produced food available to local residents. For example, six hundred families obtain their weekly provision of fresh produce from Oxbow. The farm also provides kale, delicata and summer squash, shallots, parsnips, and lettuce to a regional chain of pizza restaurants and sells produce wholesale to local restaurants, co-ops, groceries, and hunger relief partners.

The native plant greenhouse and education programs each generate about half as much revenue as they spend. The total cost of these programs is less than half that of the farm, but still, they run at a deficit. As I reflect on my experiences at Oxbow, I wonder how the rich mix of farming, education, and ecology is sustained and whether their approach could work elsewhere.

Grants written by scientists and administrators on the Oxbow staff help fund the conservation and education goals of Oxbow. The bird work that Lauren and I conducted, for example, was funded by a grant Matt received from a small nonprofit organization called Conservation, Research and Education Opportunities International. However, grants only provide minimal support for the farm. The whole enterprise is made possible by generous owners who donate money and use of the land to the project. Their investment repays them by advancing their mission to inspire "people to eat healthy, sustainably grown food and to steward our natural resources for future generations." A model, such as this, where financially successful people interested in land stewardship help underwrite a farm's operations, is one way to live in harmony with the land.

Oxbow is not the only farming operation to gain support from a benevolent benefactor. Ted Turner, who built Turner Broadcasting System, a media conglomerate with assets including CNN, HLN, and TBS, is the second largest individual landowner in North America. He ranches over fifty thousand bison across his two million acres in an ecologically sensitive way that protects land and provides high-quality food as well as increased knowledge about western ecosystems. In chapter 9 we'll meet Peggy Dulany, who is investing in a group of progressive Montana cattle ranchers that are learning to coexist with native bears and wolves. I hope these ecologically minded philanthropists will be known to future generations for their generosity in affording young farmers and ranchers a chance to succeed. Their philosophy and support are allowing us to learn how best to produce healthy food without destroying nature.

The rising cost of purchasing or leasing land is a significant barrier to young people who want to become farmers. Proprietors such as Alberg, Turner, and Dulany remove this barrier, but other ways require less of the owner. Land trusts dedicated to preserving open space that includes farmland often employ ecologically minded young farmers or lease lands at a reasonable rate. Some states, such as Nebraska, offer tax incentives to landowners who lease farmland to beginning farmers. Other states, such as Washington and Texas, lease state-owned lands to farmers at reasonable rates. And many states have land-link programs that help connect farmers transitioning out of farming with those looking to move into the trade. The International Farm Transition Network works to connect farmers and help them plan for the transfer of "money, management, and assets in a manner which allows the farm business to continue under a new primary operator." Allowing existing farms to transfer across generations provides a way for young farmers to get started and for agricultural lands to remain productive.

Measuring the production of a farm is straightforward when we consider its crop yield. But at Oxbow and elsewhere, an important output that does not contribute to the bottom line is knowledge. The lessons learned at Oxbow are shared with students, visitors, and neighbors. Seeing how the old ways of fallowing fields, rotating crops, and building soil without the use of chemicals can produce vibrant crops and native habitat surely is noticed by other farmers who may then take up these methods. Farmers change their ways when they perceive that a neighbor's crops are outproducing their own and learn from rubbing elbows with that farmer. Demonstrating how lack of tilling improves soil organic matter and crop growth while reducing input costs, for example, at Rodale's demonstration farm has converted many conventional farmers into conservation agrarians. Likewise, by inviting interested parties to tour their progressive farms in California, South Dakota, Kansas, Nigeria, and elsewhere in the world, practitioners are showing rather than telling others how to improve their bottom line through better land stewardship. The spread of knowledge through demonstration is slow—David Montgomery states in his book, for example, that despite its "rapid pace of adoption, only about 11 percent of global cropland is under conservation agriculture." In China, however, a purposeful network has increased the pace of this cultural evolution.

Chinese scientists conducted experiments across the nation to demonstrate farming approaches that small landowners could employ to increase their yields and reduce their environmental impact. More than a thousand researchers developed local recommendations concerning irrigation, fertilizer, and planting that applied to farmers in each of China's agroecological zones. What is important is that the researchers spread their knowledge throughout a national campaign network to more than 65,000 extension agents (technicians and administrative staff) and over 138,000 agribusiness personnel (production and marketing employees as well as dealers and sales reps). In this way, from 2005 to 2015 scientific knowledge found its way to nearly 21 million farmers who worked almost ninety-four million acres of land. The adoption of expertise had a stunning impact on farm productivity and environmental degradation. Average yields of corn, rice, and wheat increased by 10.8–11.5 percent, producing a net gain of thirty-three million tons of grain. As yields rose, the use of fertilizers declined by 14.7–18.1 percent. The savings from reduced use of nitrogen amounted to over $12 billion. Because the use of fertilizer was directed only where needed, the total greenhouse gas emissions were reduced by 4.6–13.2 percent.

A similar national network has been operating in the United States since the 1930s. The U.S. Department of Agriculture staffs local offices across the nation with cooperative extension agents, who distill the latest agricultural research from the region's universities and government labs into practical advice for farmers. This valued service helps farmers improve the security of their crops, build strong communities, and incorporate environmentally sound and wildlife-friendly practices into their routines.

Developing effective approaches and demonstrating their value is only one way that conservation practices spread from farm to farm. The Chinese researchers echoed what I often heard among the farmers I've met. Professor Zhenling Cui and his team concluded, "Changing farmer behavior requires more than scientifically sound and evidence-based technologies. Building trust, participatory innovation, developing human capacity and strengthening the coherence of the farming communities are critical for sustainable changes." In essence, the lessons learned stick because they are useful and the teachers are trusted. Empowering those who farm small lands to learn and share economically viable and ecologically sensitive practices is critical because 2.5 billion such farmers worldwide work 60 percent of the arable land. Oxbow may be a tiny sliver of this part of Earth, but by traveling to Central America, I learned that farmers there too are working their land and sharing their story to improve our future world.

SEVEN

Tres Amigos

A FEW HOURS AWAY FROM COSTA RICA'S Central Valley, where 70 percent of the nation's five million people reside in a dense amalgam of cities and suburbs, a two-lane highway known as Route 239 abruptly turns into a rough gravel road. As the way begins to cut hairpin turns up and over the Central Pacific Mountains, it winds between the coffee farms of Puriscal and the coastal town of Parrita. The landscape becomes increasingly rural as one descends from the height of the land through bits of jungle and pasture to Mastatal, a small village of 125 people that hugs the simple intersection of 239 with Route 318, an even smaller ruddy dirt road that leads to the indigenous community of Zapatón. Settlers founded Mastatal, named for the mastate trees (*Brosimum utile*, also called milk or cow trees) that ooze sticky white latex sap used by pioneers only a century ago as a milk substitute and analgesic. The village, carved from the wet midelevation rainforest that cloaks the surrounding mountains, hasn't changed much since this founding. Today, a primary school serves nine students, while a small library, community center, café (known as a *soda*), and bar service the farmers that live here and the would-be farmers who flock here from around the world to learn about sustainable agriculture and permaculture.

Golden ears of ripe corn are drying in the late morning sunshine of a typical autumn day in Mastatal. The corn farmer has spread his harvest on a tarp that occupies most of the village's main intersection; the daily bus has already passed, and traffic other than local motorbikes would be unusual, so the road is a nearly clear space for drying grain today. Don Chépo, the owner of the soda, keeps an eye on the crop just in case. My students and I must step around the harvest as we look for birds and hike toward a refreshing swim in the Río Negro. As we walk we move from one small family farm, called a *finca*, to the next. Planted trees border the pastures and orchards, which are worked by hand and horse. Birds are ever present: guans, doves, wrens, owls, toucans, tanagers, woodpeckers, motmots, parrots, and anis,

Opposite page: Ferruginous pygmy-owls, such as this one perched near a shade-coffee plantation, move freely between reserved lands and surrounding farms.

to mention just a few! The fincas are a stark contrast to Costa Rica's larger corporate farms of banana, pineapple, and coffee, where owners spray more pesticides and herbicides per acre than anywhere else on Earth.

ISIDRO GARCÍA DÍAZ mounts his horse, Amarillo, and descends a steep driveway to the small pasture just across the Mastatal road. His daily farm chores include rounding up the small herd of dairy cows so that his wife, Arabela Guzmán Cerdas, can milk them and make her daily two pounds of fresh cheese. Yellow-headed caracaras, scavenging hawks that are common throughout Costa Rica, ride the backs of the cows, searching for parasites among the animals' hides or insects flushed as they move about the pasture. In wilder areas, these birds ride tapirs, but they seem equally at home working cattle. The fence and rails that confine the cows were hand hewn from the local forest and now carry a rich, russet patina courtesy of the muddy Brahmas that have rubbed, bitten, and kicked at them for generations. Isidro and Arabela raise corn, tomatoes, onions, garlic, a variety of fruits, chickens, turkeys, pigs, dairy cows, and bulls on their thirty-five-acre finca. They purchase beans and rice, which are essential to the traditional diet of Costa Ricans, or acquire them from a nearby brother's farm. Their produce has nourished themselves and their four children for thirty-five years and seems destined to do so for many more, thanks in part to their eldest son, Marcos.

Marcos Alonso García Guzmán is one of three Gen X Ticos farming in Mastatal who are modernizing and carrying forth the sustainable techniques of the previous generation. (Costa Ricans fondly refer to themselves as Ticos.) I met Marcos on my first visit to Mastatal in 2009 and have enjoyed watching his agroecology program at Finca Siempre Verde grow each year since. Marcos's finca is a short walk from two others. A couple of miles up the road toward Zapatón, Javier Zúñiga runs a polyculture farm named Villas Mastatal. Two pastures down from Finca Siempre Verde, Jorge Salazar turns cacao into handmade chocolates at La Iguana Chocolate. These three friends and their families have long histories and deep roots in Mastatal. Each spent time away from the village exploring other professions, but they eventually returned to develop small-scale, organic farms. What is it about this rural crossroads that has lured these three young men home

Isidro García Díaz moves the dairy herd back to their pasture.

to care for the land? In part, the explanation lies in the creation of a new national park, La Cangreja, which stretches from the small town of Santa Rosa to Mastatal. Residents of Mastatal believe that the park might support an ecotourism industry, much as the other national parks of Costa Rica do in their gateway communities. However, the creation of the park in 2001 wasn't the only force relevant to my friends' decisions. A special couple also moved to Mastatal the same year, bringing with them an ideology of sustainability.

Tim O'Hara and Robin Nunes, Peace Corps alums from the United States, founded Rancho Mastatal, a three-hundred-acre farm, education center, and forest reserve in the heart of the village. Tim and Robin employ local workers and underwrite the construction of community projects, such as the library. But their lasting influence comes from the sustainable living solutions that they model on the Rancho. Local people as well as interns and workshop attendees from around the world come to the Rancho to learn about building with local materials, composting, biogas generation, and organic agriculture. Marcos, Jorge, and Javier saw opportunity in the way Tim and Robin established an agrotourism business and how they worked the land. And so, they set forth to blend this new approach with the traditional workings on their parents' property to sustain the agricultural lifestyle of Mastatal.

Isidro initially resisted Marcos's idea to expand the farm. In the elder's mind, using the forest to shade his cattle was sufficient. Perhaps he also wanted his son, a certified architect and carpenter who had a good command of English, to find a better life off the farm. But Marcos, then twenty-two years old, persisted and finally convinced his papa to let him build a bunkhouse and classroom that would start Finca Siempre Verde. The venture has been wildly successful since its meager start in 2005. Tropical fruits, such as bananas and starfruit, now hang from the trees that shade the buildings constructed from local wood and cob—a strong mix of clay, sand, and hay that is a sustainable alternative to concrete. A simple greenhouse shelters lettuces, tomatoes, and peppers, while herbs, such as lemongrass and wild cilantro, grow in tidy outside gardens. A living fence of almond trees feeds the scarlet macaws—huge noisy parrots plumed in red, blue, and yellow that live throughout Costa Rica's western forests—and shelters the bananas from damaging winds. A small cadre of four or five pigs provides meat, fertilizer, and fuel. A large rubber bladder, lying prone like a huge sausage, expands as it captures the methane emitted from composting pig poop. A simple copper pipe directs the methane to fuel the kitchen stove. Chickens also help recycle farm waste into meat, eggs, and compost.

To visit Finca Siempre Verde is to immerse oneself in sustainability. This evening I eat a splendid meal of salad, fruit, and chicken, all from the farm. Here, farm products provide 40 to 60 percent of foods consumed by visitors and family each year. The chicken was cooked using compost-generated methane. The tropical sun warmed my evening shower of captured rain. And my waste was recycled in a composting toilet, constructed with only three sides and set to provide a fantastic view of a bromeliad-studded oak tree that has a dead limb replete with a cavity frequented by a yellow-naped woodpecker. Siempre Verde is the first place I've visited where I honestly felt connected to the nutrient and energy flux of an ecosystem: as I ate the land's bounty my body extracted the sugars, minerals,

An adult (right) and juvenile yellow-headed caracara search for insects among Isidro Díaz's cattle.

and vitamins it required and returned the rest to amend the soil and nourish the next crop.

I am not alone in appreciating the sustainable farm life. As I enjoy an afternoon view of the nearby mountains from Marcos's deck—the peaks of La Cangreja rise to four thousand feet in the north—a trio of fiery-billed aracaris pauses in the fig tree, not ten yards away. These beautifully painted, smallish toucans immediately attract my attention. Their serrated beaks can pluck large insects and ripe fruits as needed from the forest. But soon I'm distracted by a flock of smaller birds working the canopy just above me. I can identify shining honeycreepers, speckled tanagers, palm tanagers, and tropical gnatcatchers before the flock is gone. As rain clouds build I watch a party of groove-billed anis hunt the shade trees and listen to the bubbly duet of a pair of rufous-naped wrens that are adding a few dried leaves to their nest in the top of a nearby cecropia tree. I can hear the rain approaching from the southeast. As it pounds the distant forest canopy, it sounds like the ocean's roar. The mountaintops disappear into gray, while wisps

The mountains of La Cangreja National Park, as seen from Finca Siempre Verde, are spared from local agriculture.

of smokelike clouds rise from the lower incised valleys. The scene is a harmony of gray, green, and yellow.

Darkness comes quickly in the tropics, and a wild diversity of insects, frogs, toads, snakes, bats, and small mammals rustle about under its protective cloak. We take a short night hike and spot a common opossum, a woolly opossum, and a vine snake on the farm. A dry forest toad hunts the kitchen floor for insects attracted to the small light. Here, even inside, there are no barriers to nature's rhythms. Marcos and Isidro share the entirety of their farm with wildlife.

Sleeping late into the morning is impossible for a birder in the tropics. My slumber is interrupted during the night by the rising and falling whistles of common pauraques. At 5:00 a.m., the constant call of the ferruginous pygmy-owl, which reminds me of a truck's backup siren, is finally mellowed by the single tremulous call of the little tinamou. In the next twenty-three minutes, I note the calls of rufous-naped wrens, blue-crowned motmots, cocoa woodcreepers, great kiskadees, scarlet-rumped tanagers, great tinamous, and red-billed pigeons. This concert of native avians rises above the din of dogs, roosters, and motorcycles as a pair of guacos, or laughing falcons, joins the gig at 5:23 a.m. and calls without a break for three minutes: *waa, waaah, co, waaahco, waaahco . . . .* While the falcons laugh, a flock of orange-chinned parakeets chatter by, and a troop of groove-billed anis noisily awaken. The calls of white-crowned parrots, fiery-billed aracaris, crested guans, and melodious blackbirds round out the dawn chorus, which finally winds down around 5:45 a.m.

A short bird walk this morning reveals two species I've never before seen. Despite this being my seventh year visiting Mastatal, today I get my first glimpse of a yellow-crowned tyrannulet and a stripe-headed sparrow. The sparrow works the brushy sides of Mastatal's red clay and gravel road, while the tyrannulet (a small flycatcher) sallies forth from a bare branch to pick insects from the sky. In three days of birding, I sight fifty-two species. That is barely one-sixth of the more than three hundred species annually counted in and around La Cangreja. The incredible diversity of birds here and globally in mountainous regions results from rapid and sustained rates of speciation at high elevations due mostly to isolation and variety (in climate and resulting ecology) that characterize complex topographies. Tropical mountains get an extra boost to diversity because of the consistent energy afforded to plants by warm, moist conditions. Greater plant production translates into increased food for animals, which allows more distinction and less extinction of the biota. Primarily, the diversity that is created by the uniqueness of mountain life is maintained in the tropics by long-term environmental stability. Stunning variety makes it hard to pick a favorite from the many hummingbirds, trogons, flycatchers, waterfowl, parrots, tanagers, and hawks. But watching a blue-crowned motmot dust bathe in the red dirt next to Isidro and Arabela's baby-blue house is a Kodachrome moment I won't soon forget.

Finca Siempre Verde sustains the local ecology and culture, but what about economics? Marcos shares his business venture with his partner, Jennifer de las Angeles Mora Pérez. Jenny, the eldest in a family of four that grew up in San Miguel, a forty-minute walk up the road, helped to manage programs at Rancho Mastatal for two years, putting her university degree in sustainable tourism to work. Together, Marcos and Jenny lead nature tours and host interns, volunteers, and students who want to learn Spanish

The fiery-billed aracari's beak is serrated to better hold insects, fruit, and nuts.

and sustainable practices in building and agriculture. They also facilitate visits by university groups and others looking for a unique retreat. Marcos and Jenny have a small house with a communal kitchen that they share with the farm's visitors. They support their visitors with a dormitory that sleeps eighteen people and features a large yoga deck. When fully booked, the farm's two showers and two composting toilets are busy but adequate. In the first five years of their business, they provided training to more than fifteen hundred students, some of whom pay to learn and others who work in exchange for experience. Their annual income, though modest by U.S. standards at $20,000, fully supports them and contributes to the needs of Marcos's parents and two younger siblings. The farm provides approximately half of their income, while the other half comes from leading nature tours throughout the southern Pacific region of Costa Rica. The economy of Finca Siempre Verde sustains simple and fulfilling lives that are enriched by international friendships forged in the red tropical soils of the farm.

CACAO ONCE SUSTAINED the community of Mastatal. Farmers formed a cooperative to gain an economy of scale in managing 370 acres of cacao plantations that included shared ownership of a fermenting facility to process the raw beans. Juan Luis Salazar joined the co-op and planted cacao on his farm in 1985. He enjoyed several years of strong production until the early 1990s, when a resurgence of the fungus *Moniliophthora roreri,* which produces the disease known as frosty pod rot that decimated Costa Rican cacao farms in the 1970s and 1980s, forced most local farmers to cut their cacao trees and plant pasture grass for cattle instead. Juan Luis did not cut his plantation, but he did stop the harvest, as it was economically not sustainable. In 2006, at the urging of his son Jorge, the Salazars opened a small artisanal chocolate company, La Iguana Chocolate, and once again harvested cacao pods from a small portion of their orchard. In so doing they bucked the national trend where abandoned cacao farms were bought up by corporations and turned into monocultures of bananas and oil palm. This homogenization of the landscape reduces biodiversity and food security, but neither is the case at La Iguana Chocolate.

White-tipped doves, social flycatchers, and red-lored parrots coo and chatter from the rich mix of trees that greet visitors to the Salazars' chocolate factory. Cacao trees dominate the twenty-two-acre farm, but this is no pure monoculture. One-third of the property retains a cloak of native forest, cacao covers ten acres, and five acres grow a combination of fruit trees and cacao. Wanting to sustain both soil fertility and provide food for the family table, Jorge and his father have diversified Finca La Iguana's cacao orchard with native *Inga* trees that fix atmospheric nitrogen into a natural soil fertilizer, tropical cedar trees, whose leaves provide a water-conserving cloak of mulch at the end of each rainy season, and fruits, such as banana, orange, and coconut. The cedar mulch is especially important; it not only enriches the soil with organic matter but by keeping the ground moist longer each year extends the cacao flowering season, enabling each of their 5,500 trees to produce 30 to 150 fruits annually. Cedar, cacao, turmeric, and vanilla are cash crops for the farm. Fruits, vegetables, and chickens help feed the family, which includes Jorge, his parents and grandmother, his girlfriend, and his younger sister.

At twenty-eight, Jorge is the youngest of the three farmers I've come to know in Mastatal. Before carving out his niche in the chocolate trade, he picked and planted pineapples in the lowlands. Jorge laughs when I ask him if he misses the back-breaking work. He tells me it was quite the opposite; he missed the tranquil lifestyle of the family farm when he commuted to work near the beach. Still, Mastatal is small, and so Jorge enjoys a couple of weeks each year away from the village in England and Belgium learning more about the art of making fine, single-estate chocolate. His expertise in arboriculture, as well as his experience making and marketing chocolate in a tropical climate, has created other opportunities to travel, share, and learn as a consultant. He augments his farm income a bit by working for a translation service. To broaden their perspective, Jorge and his sister took advantage of a national learning program that coupled home study with testing in the nearby town of Puriscal to earn university degrees in sustainable ecotourism. Schooling gave them the skills needed to open the farm to interns and visitors, increasing profitability (about half of the farm's earnings come from tourism, while half come from chocolate sales) and enriching the farm scene for young farmers eager to meet, befriend, and learn from others. The spark of international exchange is evident as Jorge explains to my University of Washington students how cacao becomes cocoa.

Chewing a few nibs of roasted cacao kernel, Jorge swings a machete, surgeonlike, to cut a ripe yellow cacao pod from a tree's trunk and carefully slice the cucumberish fruit crosswise to reveal the slippery white pulp that envelops forty to sixty seeds. He passes the fruit among the students telling them to pop one of the slick beans into their mouth and savor the sweet, thirst-quenching sensation of the pulp. Apprehensive initially, we quickly discover firsthand why monkeys, squirrels, and coatis routinely raid Jorge's orchard, taking up to sixty-six pounds of produce each week. Though the pulp tastes nothing like chocolate, we could quickly get hooked on this unique food.

It is the seed, not the pulp, that has Jorge's attention. Tending the seeds is a low-tech, mostly hand-powered process at La Iguana. Seeds extracted from the pod are placed outside in wooden boxes for about a week to ferment. Jorge turns the brewing seeds each day with a simple wooden shovel. Fermentation converts the pulp to alcohol and mellows the bitter taste of the seed into the rich chocolate flavor we crave. The breakdown of pulp generates considerable heat, which today has the mash at 110° F —

Cacao seeds drying at La Iguana Chocolate.

about twenty degrees above ambient. This high heat doesn't discourage flies and bees from crawling into the beans and sipping the alcohol. (This cacao wine, or *chicha*, is like kombucha, but not very popular. The resulting vinegar, however, is sometimes used on salads.) It might seem unsanitary to have insects walking among the seeds, but Jorge welcomes these ecosystem servers as they help inoculate the beans with wild yeasts, the active ingredient in the fermentation process. When adequately fermented, the cacao beans are dried on screens in a plastic-covered greenhouse for a week or more, depending on humidity. After all of this, the cacao is finally ready to be roasted into cocoa.

When I visited Jorge eight years ago, he roasted the beans in a large iron kettle over a wood fire or used pedal power from a jury-rigged old bicycle to turn a roasting basket over the fire. That crude process only produced smoky-tasting chocolate. Today, he is a bit more modern, using an electric skillet that more precisely controls the temperature and allows him to bring out the beans' other flavors and produce a greater variety of candies. In twenty-five minutes the seed shells are crisp and brown and ready to crack open. We relish the deep and familiar aroma of chocolate as we roll each seed between our fingers to crack the shell and extract the roasted kernel, or nib. Each nib produces about a gram of chocolate if we don't eat it directly. We grind the nibs that we spare with stones cranked by hand to extract cocoa butter and produce grainy chocolate. We make various simple candies by combining locally processed raw cane sugar called *tapa dolce,* cocoa butter, and ground nibs with flavorings, such as rum, almond, mint, and vanilla. After a few minutes of cooling, we gorge on our delicacies and the chocolate milk and brownies the Salazars serve. We are all a bit buzzed by the high cocoa content of our feast, but hey, dark chocolate, especially that with over 70 percent cocoa, like the palm-oil-free goodies we've eaten, is healthy as well as tasty.

As we come down from our chocolate high, we learn more about the diversity of products the Salazars make from cacao. Candies, lotions, balms, butter, powder for baking, and nibs for the real addicts are packaged and advertised locally and through social media. In total each year the family moves about 2,200 pounds of cocoa. That effort nets three dollars for each 2.2 pounds of roasted beans sold or fifteen dollars for each 2.2 pounds of product sold. The farm's offerings are continually evolving. Attractively wrapped fine chocolate bars are the newest high-end product added to the La Iguana Chocolate line. Jorge and his girlfriend, Vicky, recently learned the tempering techniques necessary to harden these bars from Belgian chocolatiers who visited the farm. In conjunction with income from farm tours and a few side jobs, the business wholly sustains six family members.

Jorge is committed to growing and continuing the modernization of La Iguana, but he is wary of getting too big and losing his ability to manage the whole process from cacao bean to chocolate bar. The farm is currently using about 80 percent of what they can produce, so without buying or renting more land or cooperating with other farmers to harvest more trees, La Iguana is nearing capacity. Jorge sees direct trade from the farm to the consumer as his most profitable option. Seeking certifications such as fair trade or organic may broaden his clientele, but they also bite into his profit. At this point, he isn't sure if he will pursue those options. His direct sales of roasted beans and bars are increasing as new customers, such as a beer company that is brewing his beans into its chocolate stout, are finding out about La Iguana.

What strikes me most at La Iguana is the natural expression of Aldo Leopold's land ethic. Jorge and his family fully appreciate the community that shares their land, as well as the commodity that comes from it. Rather than killing the wild animals that live on the farm and eat a substantial number of cacao pods, they foster a more diverse wildlife community so that the numbers of challenging animals are kept in check naturally. Giant snakes, such as boas, are valued here because they eat squirrels. Hawks and eagles are encouraged because they reduce rats and mice. Monkeys are harassed out of the orchard rather than killed so that they learn where they can and cannot tread. A small pond and wetland adorn the property and help introduce visitors to the unique birdlife of the area. A little jet-black bird vanishes under the surface as I work my binoculars along the edge of the pond. Soon it pops up and stares at me with a piercing yellow eye. The least grebe I'm watching is a testament to Jorge's efforts to share the land with wildfowl. I observe it fish and float for nearly half an hour before ticking it off my list—another new species from the seemingly endless bird diversity of the neotropics.

THE THIRD ORGANIC family farm in Mastatal is an experiment in restoration ecology, agroforestry, and ecotourism. Javier Zúñiga, his wife, Raquel, their two young children, and Raquel's parents, Mario and Lucia, live on and work the fifty acres Raquel's family has farmed for most of the past century. Coffee was king of the farm for sixty years, but when prices fell in the 1980s, the family replaced coffee with pasture, and cattle pounded the land until 2006, when Javier and Raquel wanted to take the farm in a more sustainable direction. The fulfillment of this dream, a work still in progress, is Villas Mastatal.

Javier, the elder amigo at age thirty-six, smiles broadly as he shows me the forest of more than six thousand trees that he, Raquel, and interns have planted in the past decade. The conservation forest, as he calls it, was a twenty-five-acre pasture that now stabilizes a steep slope and provides an amazing array of foods and other products. This food forest has a low canopy of nut and fruit trees, such as cashew, citrus, cacao, coffee, almond, banana, papaya, pineapple, mango, mamón chino (rambutan), moringa (a multiuse tree with edible pods, leaves, seeds, and roots rich in vitamins and minerals), and avocado. The forest also includes melina trees, which provide all the wood and timber required for building and accompany native botarrama, caobilla, and cristobal trees in rising above the edible middle canopy. Terracing the ground by digging parallel swales increases the area for crops. The upslope of the swale is stabilized with deep-rooted vetiver grass, while downslope grow yuca (a local sweet-potato-like tuber known elsewhere in the tropical world as manioc or cassava), beans, squash, and sugar cane. A greenhouse sprouts lettuces and nurtures seedlings of the rare tree *Plinia puriscalensis,* the fruits of which grow from the trunk. When they are large enough, Javier transplants the *Plinia* to sensitive areas in local restoration projects. The remaining half of the farm has pasture for the livestock that provide meat, eggs, milk, and cheese, along with housing for the Zúñigas and their visitors. The Zúñigas' goal is to make an ecologically sensitive retreat for international visitors and vacationing Costa Ricans. They cater to a wide range of tourists by providing three levels of accommodations: a dorm for students seeking inexpensive lodging, rustic cabins for vacationers, and a high-end villa for tourists looking to be pampered. The farm's forests provided all the lumber for the buildings, and the 150 visitors per year are fed nearly exclusively food produced at Villas Mastatal.

The diversity of life in the dense, wet forests of the neighboring national park seems to flow smoothly across the Villas. We glimpse the abundant birdlife of this unique farm as a collared forest falcon rushes through the conservation forest, interrupting Javier's lecture. The eerie calls of great tinamous resonate from the deeper forest below us, trogons, tanagers, and euphonias work the fruit and nut trees, while white-collared and spot-breasted swifts, orange-chinned parakeets, and black vultures fly through a magnificent vista centered on the high peaks of La Cangreja. Careful use of resources—recycling gray water from cabins to laundry and finally to toilets and composting plant matter, human waste, and cow manure—blurs the line between the working land of a farm and a nation's treasured reserve just across the narrow gravel road. Diversifying the corners and borders of the farm, for example building a small pond for native frogs, toads, and waterfowl, adds unique wildlife habitat to the area.

The wildlife lures birdwatchers, such as me, to Javier's farm. Returning to the farm in 2017, I meet Alejandro Guzmán Chacón, a guide who is staying in the cabins with three Ticos. They've come from Puriscal for a weekend of birding and find the farm the perfect base because of its proximity to La Cangreja. Alejandro is delighted that so many of the park's birds can be seen on the farm. He sees little difference between the conserved forest across the road and the food forest, where we sit drinking its coffee and eating empanadas featuring its cheese and sugar. His farm list includes forest species such as the black-throated trogon, turquoise cotinga, and blue-crowned manakin. On open ground, his customers and he appreciated the subtle dance of a male blue-black grassquit as it jumped several feet off the ground and sang a buzzing trill. The many birds we enjoy watching also help sustain this finca. Hummingbirds flock to the nectar in the forest and garden. In so doing they pollinate flowers. Tanagers and toucans are attracted to the cecropia trees and lingua-de-vaca shrubs seeking fruit and insects. Their appetites negate the need for insecticides.

Providing the ecological and laid-back setting that some tourists crave seems a natural fit for Villas Mastatal. A visitor from California joins us on the porch; she's enjoying the tranquillity and power of a late afternoon downpour. Changing the focus of the business from farm tourism, where interns and short-term visitors from organizations such as WWOOF and Workaway come to help with farm chores, to ecotourism, where visitors pay a fair price for immersion in nature and local culture, makes economic sense to Jorge. He notes that increasing opportunities for farm tourism elsewhere in Latin America and around the world has reduced interest in Villas Mastatal and that the costs of supporting visitors often exceed the benefits of the work they accomplish. By focusing on ecotourism, Jorge can increase income and employ local workers to do the jobs farm tourists once did.

THE CROWING OF ROOSTERS blends with the cackling of gray-cowled wood rails as the sun dawns on Mastatal. A droning motorbike breaks the avian duet when a young man drops a student at school before commuting to a distant job. This is the daily hustle and bustle of a small farm community that seems mostly content with the simple life. Some of those raised here leave to build urban lives. Two of Marcos's siblings, for example, teach and practice law in the Central Valley. But most remain close to the red clay and dripping green forest that has sustained their families for three generations. The farms tended and nurtured by Marcos, Jorge, and Javier give me hope that life in harmony with the land will continue to exemplify this small tropical crossroads. Together the three farms we've visited support 12 percent of the Mastatal population. The similarities in services offered by these three entrepreneurs have led to some competition for tourists and therefore to changing emphases at each farm. As Javier and Raquel build up their lodging to serve a greater number and diversity of visitors, Marcos and Jenny provide more nature guiding services, and Jorge seeks direct access to distant markets for his chocolate. The diversity of services now offered in Mastatal increases the village's resilience to changing markets driven by world events. Each family is successfully mixing tourism and education with small-scale, organic food production. In so doing they foster local conservation, build appreciation for nature among their many visitors, and support their extended families on land previously farmed by their fathers and grandfathers.

In contrast to the North American croplands, where the number of farms and farmers is shrinking, here in rural Central America, I've found stability in the number of both farms and farmers. In Mastatal, Generation X is remaining on the farm their parents and grandparents homesteaded. This seems possible for several reasons. First, the young farmers I've met all prefer the tranquillity of rural Costa Rica to the rush of the city. Second, a commitment by the government of Costa Rica to reserve almost a third of the country in national parks has juxtaposed natural wonders with agricultural land, thereby providing ecotourism opportunities for landowners. Third, and most important, the aging patriarchs of each farm were willing — with some persistent convincing by their sons and daughters — to allow their children to change the complexion of their farms. I'm sure Juan Luis and Isidro never imagined that they would host hundreds of visitors each year who wanted to experience Costa Rican farm life for themselves.

Indeed, they could not have believed that these visitors would help their children expand the farm's income to support a growing family. Whereas the last generation farmed out of necessity, the current generation farms and teaches out of concern for the environment, from a deep love of country life, and because of the international business opportunities it affords them.

Although the three farms I've visited in Mastatal share a vision for the future of their region, they act mostly independently. An exception was their cooperation in 2014 to found a nationally registered nonprofit organization called the Ecoemprendedores. These ecovisionaries can lobby the government on behalf of the national park and the village of Mastatal. This has helped promote Mastatal as a touristic community, provided trail maintenance and interpretation in La Cangreja, and increased the residents' access to language classes, natural history training, and other social activities. For example, each year the group teams up with the national park service to organize a bird count and host a run. These events draw people to Mastatal for the mutual benefit of all farms and businesses in the village. Proceeds from the annual run also return to the community, for example, to fund new goals, a mower, and nets for the community soccer pitch. Mastatal is not alone in benefitting from the Ecoemprendedores; members of the group travel to other parts of Costa Rica to discuss their success in community-based conservation and ecotourism.

THE CONTRAST BETWEEN small organic farms and large, intensively worked farms is stark as one travels to Mastatal from either Puriscal or Parrita. In and around Puriscal acre upon acre of coffee engulfs the midelevation slopes of the mountains that form a high, treed spine down the center of the nation. Coffee is Costa Rica's third largest cash crop (bananas and pineapples rank first and second, respectively). In 2016, the nation's farmers exported over three hundred million dollars' worth of coffee. The affluence this provides is evident as one walks the streets of coffee towns such as Puriscal or Santa Maria de Dota: trucks are new, streets and sidewalks well maintained, houses constructed of tidy masonry, and schools first rate (though less attended because of the lure of high wages to be made in the fields). The regularly spaced bushes, shorter than the average person, are

Native trees create a broken canopy above shade-grown coffee bushes, providing habitat for birds.

dense with coffee cherries but offer little food and shelter for native wildlife, unless taller trees are studded throughout the crop.

Orchards that mix trees, such as banana and *Inga,* provide natural subsidies to the coffee crop and offer habitat to wildlife. If such trees cover 40 percent of the crop in Costa Rica, they can be certified as shade grown. If the owners also embrace organic practices that shun chemical fertilizers and pesticides; grow a mixture of native ground covers and trees that are allowed to accumulate epiphytes and occupy various positions in the food forest, including some trees that reach to heights of at least forty to fifty feet; and use native vegetation to delineate tree lines and buffer streams, they may also be certified as bird friendly by the Smithsonian Institution.

Studies abound on the benefits of shade-grown coffee, especially when the plantation begins to resemble a complex forest, as do those considered bird friendly. A 2006 review of more than fifty research projects found many demonstrations of the ecological advantages of shade- versus sun-grown coffee. Besides increasing the crop's sale price, improving soil health, supporting pollinators, reducing insect pests, and increasing carbon storage, shade-grown coffee provides important habitats for migratory and resident birds. Although birds requiring deep forest avoid coffee plantations of all kinds, those that live in more open conditions are more abundant, more diverse, and in better physical health when they inhabit shade-grown rather than sun-grown coffee farms.

After a morning cup of java, I bird the shade-grown coffee fields owned by the Coopedota farming cooperative south of Puriscal. Hoffman's woodpeckers and blue-gray tanagers visit the bromeliad-laden *Inga* trees. A small group of white-tipped doves forage in the open ground, while a white-naped brush-finch, a Wilson's warbler, and a flock of sooty-capped chlorospingus work the coffee bushes. A family of groove-billed anis lures me to the hedge bordering the orchard, where I spot a tiny brown-and-white Costa Rican pygmy-owl ignoring the protests of tropical kingbirds and mountain elaenias. The stream leaving the farm hosts a loudly singing riverside wren and empties into an adjacent cattle pasture alive with the whistles and songs of eastern meadowlarks, great-tailed grackles, black phoebes, and small *Empidonax* flycatchers. Unfortunately, the vast majority of coffee farms in Costa Rica are much simpler, high-yielding monocultures that rely on inputs of agrochemicals, rather than natural compost, nitrogen, and shade provided by a mix of trees and coffee bushes. A few years earlier, in 2014, my graduate student Avery Meeker and I experienced the difference. We discovered twenty species of birds in a shade coffee farm but only twelve in a nearby sun coffee plantation.

Descending from the coffee farms along the southern Pacific coast, African oil palm plantations have replaced mainly the once abundant banana plantations farmed by the United Fruit Company. Global demand for cheap oil that is now found in "half of all products on U.S. grocery store shelves — from crackers and ice creams to lotions and lipsticks," according to journalist Jocelyn Zuckerman, has devastated tropical forests throughout Asia, and I wonder if it will have similar impacts in Costa Rica. Worldwide demand for cheap cooking oil and government incentives to convert palm oil to biodiesel spurred a near doubling of the area covered by oil palm plantations from 2003 to 2013. Although most of this expansion has

The still and dark interior of a twenty-five-year-old, mature oil palm plantation.

occurred in Malaysia and Indonesia, the land devoted to oil palm in Latin America has steadily increased as well (at an annual rate of 9 percent from 2001 to 2014). When plantations replace tropical forests, as is typical in Southeast Asia and South America, the loss of biological diversity is staggering. In plantations, the variety of ant, frog, bat, beetle, bird, butterfly, moth, small mammal, and primate species is less than half of what it is in nearby forests, even those previously logged several times. Clearing forests for oil palm endangers orangutans and Asian elephants. Forest birds, such as woodpeckers, those reliant on native fruits, such as helmeted hornbills, and those requiring understory vegetation, such as pheasants and pittas, are typically absent from oil palm plantations. Conversion of lush tropical forests to palms also dramatically reduces the ability of tropical lands to sequester carbon and buffer climate change.

Costa Rica is a relatively small player in the global palm oil market, and because its oil palm plantations are sited on lands formerly cleared for agriculture, their impact on local ecosystems appears to be less consequential than when palms replace forest. Throughout Latin America, 98 percent of oil palm plantation expansion from 1989 to 2013 occurred on former agricultural lands. A 2014 survey found that palm expansion in Costa Rica occurred primarily on cattle pastures, croplands, banana plantations, and forestry plantations, such as those producing timber from teak. Costa Rican law prohibits clearing of native forests, which results in the redevelopment of ag lands for oil palm to a greater extent than in nearby Ecuador and Peru, where rainforest clearing provided sites for 40 to 70 percent of new plantations.

Growing trees where grassy cattle pastures once stood may benefit birds and improve a patch of land's ability to sequester carbon, but to me the plantations are reminiscent of the lonely cornfields of Nebraska—row upon neat row of gray or brown trunks with green fronds only sprouting from the top, like a wild hairdo. I hear that the aggressive and venomous viper locally called the *terciopelo,* or fer-de-lance (*Bothrops asper*), is abundant in the rat-infested understory where decadent fronds are piled high. As we drive by miles of palm, I see a few vultures and roadside hawks. Vultures are known to eat the fruits of oil palm, and raptors often hunt palm plantations for rats. I also often spy white ibis and cattle egrets wading the small, nutrient-rich streams that collect runoff, suggesting that here, as in Malaysia, these habitats support some native birds. It was time to step inside a plantation and get a more accurate count, so after we birded various coffee plantations, Avery Meeker and I walked into the shade of a oil palm plantation outside of Parrita. We tallied seventeen different birds, seven of which inhabit streams and standing water. House wrens and variable seedeaters foraged among the decaying piles of fronds. Only six species used the trees, and mostly merely for perching (roadside hawk, great kiskadee, great-tailed grackle, smooth-billed ani, and streaked flycatcher). However, the Hoffman's woodpecker appears to nest within the plantation: we heard the birds calling and noticed several nests or roost holes in the few standing dead palms.

If working the land hard, as on palm plantations or sun coffee farms, spares other areas, then conservation may be well served. This approach hinges on the assumption that, as fewer people are needed to farm the intensive, technologically advanced farms of the future, rural lands will transition back to a native state. This shift has occurred in eastern North America because the wealth initially produced from agriculture enabled local industries to develop, and the secure, high-paying jobs from industry drew people from the countryside, allowing forests to reclaim old pastures and fields. However, this is far from an absolute or lasting fate. Ecologists Ivette Perfecto and John Vandermeer, from the University of Michigan, contend that sociopolitical and ecological conditions determine whether technological progress either results in land sparing, as farmers can produce more on less acreage, or exacerbates consumption of land, as increased profits lead farmers to expand production. The westward spread of corn in Nebraska, for example, illustrated the vulnerability of spared private lands to changing crop prices.

Professors Perfecto and Vandermeer argue that, in the neotropics, agricultural intensification will most likely lead to agricultural expansion rather than contraction. They believe this in large part because the multinational corporations that operate in the tropics see tremendous opportunity in the expansion of industrial agriculture, which employs low-cost labor and exports high-profit foodstuffs. Because of this possibility, Perfecto and Vandermeer "challenge the assumption that agriculture is the enemy" of conservation, noting that "it is the kind of agriculture, not the simple fact of its existence, that matters." The kind of agriculture that provides wildlife habitat, and therefore aids conservation, is neither intensive nor industrial but instead is the small-scale, mixed-crop approach that I encountered in Mastatal.

These farms support a significant fraction of the avian tree of life. As I cataloged birds in 2016, I noted representatives of all the major branches of the bird phylogeny. I saw primitive tinamous, whistling ducks, grebes, guans, tiger-herons, sunbitterns, vultures, and hawks. Each day I encountered many types of the more advanced rails, pigeons, parrots, anis, owls, nighthawks, potoos, swifts, hummingbirds, trogons, kingfishers, toucans, woodpeckers, woodcreepers, and antbirds. Both on and off the farm, I enjoyed dozens of the most recently evolved species: the falcons, flycatchers, vireos, jays, swallows, wrens, thrushes, warblers, tanagers, finches, buntings, sparrows, orioles, and blackbirds. My experience is not unusual.

Professor Gretchen Daily and her colleagues at Stanford University studied Costa Rican birds on farms and in forest reserves for a dozen years. They found representatives of all the branches in the avian tree of life represented on farms. But they also found striking differences in the richness, ability to preserve evolutionary distinctiveness, and likelihood of extinction on farms versus reserves. Small farms, such as those in Mastatal that grew a diversity of crops—typically around thirty species of plants—held about the same number of bird species as did forest reserves. Both reserves and diverse farms held more types of birds than did intensively farmed monocultures of oil palm, banana, coffee, or pineapple.

Notwithstanding the ability of mixed agricultural lands to maintain the diversity of birds, they supported 15 percent less of the avian evolutionary history found in reserves. Despite an ability to harbor many species, that drop occurred because birds on mixed farmlands were more closely related to one another than were the birds found in forest reserves. Farm monocultures also featured mostly closely related birds, but because these intensive agricultural lands also held the least diversity, they supported 40 percent less evolutionary history than did reserves. Most species able to thrive in monocultures are recently and rapidly evolving bird groups, such as sparrows and blackbirds. These species are expected to survive into the future even as intensively farmed monocultures continue to extirpate more evolutionarily distinct species.

The findings of the Stanford study and my own experiences on the mixed agricultural lands of Mastatal suggest that resisting the intensification of agricultural processes has immediate and lasting benefits for a beautiful diversity of tropical birds. Diverse agrarian lands cannot provide what all species require, but they are much better at this than are nearby intensively farmed monocultures. Not only do such lands conserve diversity themselves, but they help prevent extinctions in nearby reserves, such as La Cangreja. Expansive, intensively worked oil palm, sun-grown coffee, or cattle pasture inhibit movements of forest animals, whereas retaining some natural elements on farms facilitates movement. With little or no movement, dispersers from afar cannot resuscitate a species after natural dips in its population size. Such was the case in Ecuador where pasture and oil palm surrounded a forest reserve. In the first five years following the Rio Palenque Reserve's isolation, ornithologist Charles Leck documented the loss of 26 percent of its birds and mammals.

Laboring on the intensive farms around a reserve erodes human capital just as it limits the land's natural wealth. Farmers working in monocultures of palm, fruit, or coffee would surely smile less than those in Mastatal today who enjoy tranquillity, grow their food, and share their experiences with a curious world. Jenny and her father are a case in point. Jeronimo Morra, Jenny's dad, is a *corbatero* on an industrial banana plantation. He works six days a week affixing insecticidal bands on the fruit's raceme to keep insects from crawling into the bananas. His day begins at 2:30 a.m. with a fifteen-mile commute. He works 7.5 acres each day wearing a suit that protects him from chemical exposure. The suit seems inadequate when spray from a crop duster flying overhead soaks him. But the sight of dead bats and birds reinforces the importance of the suit and the shower he takes to end each day in the fields. He is tired and has a bad back from manual labor, but the $2.50 per hour wage is needed, and the options for other work are limited. Jenny wishes her dad could enjoy farming as much as she does.

In naturally rich regions, such as the tropics, small farms may be especially sustainable socially and economically if they are near more substantial forest reserves. The presence of 7.2 square miles of mid- and upper-elevation rainforest in La Cangreja National Park, for example, has enabled the family farmers of Mastatal to double their income by catering to tourists as well as crops. A region that provides reserves and various farms can support the full array of birdlife. And this does not have to come at the expense of lowered food production. Small-scale farms that tune their crops to the local ecological conditions are quite productive. The yields of organic farms are typically 20 percent lower than those of conventional agriculture. However, when organic farms are diversified through crop rotation and the simultaneous growing of multiple crops within a field, they approach to within 8 percent of the conventional yields. Moreover, given the premium prices paid for organic produce, organic agriculture was recently discovered to be 22 to 35 percent more profitable than conventional agriculture.

As Professor Daily and her coauthors wrote, "Shepherding biodiversity through the human pressures of the twenty-first century will require a shared vision for conservation and agriculture, one that simultaneously preserves species and ecosystem functions while also enhancing food production and human well-being."

As I got to know the farmers of Mastatal, I found within them the shared vision of which the ecologists from Stanford write. To Jorge, Marcos, and Javier, considering ecology, food production, and their family well-being as interdependent aspects of farming is not academic, it is their natural tendency. If any of these three aspects lagged, their small family farms would collapse. As each year passed, I've been pleased to see my amigos' families strengthen their relationship to the land and watch as their vigorous farms and businesses repaid their respect with the economic and therapeutic means to prosper, learn, and travel. And to smile.

EIGHT

A Farm in the Wilderness

RAUCOUS SCREAMS OF SCARLET MACAWS and deep hoots of howler monkeys reverberate across ancient forests. I am hiking with students on a narrow path from La Leona to La Sirena, a ranger station deep in the wilderness that is Costa Rica's Corcovado National Park. I've left the farmers of Mastatal and traveled most of a day southwest to the Osa Peninsula, one of the world's true hotspots of biological diversity. Here among a riot of birds, bugs, and beasts, I have an opportunity to compare the wildlife on land shared with farms to that on lands spared from agriculture.

It only takes a few steps into the park to appreciate the importance of setting some land aside from human action. As we hike, suddenly the forest canopy is alive with alarmed voices as a troop of white-faced capuchin monkeys scolds a puma concealed in the thick underbrush. We never see the cat, but our adrenalin pumps as we interpret the simian voices and survey the scene. A few minutes later an oncoming group of hikers fills in the details, describing how the cat bounded across the path, just ahead of us. The possibility of spotting a big cat—a puma, jaguar, or ocelot—on the eighteen-mile trek keeps our eyes and ears on full alert as we walk. We scan the sand and soft muds for animal tracks and in short time find the unusual three-toed spoor of a Baird's tapir. This odd relative of the horse tips the scales at over five hundred pounds, yet its footprints sink only a few inches in the wet sand. Every sight and sound gives us pause and builds appreciation for land spared from the plow.

A short trail leads from La Sirena, where my students and I spend three nights, to the mouth of the Sirena River. We start and end each day with a round-trip journey between the station and the river. I've hiked here in three prior years, yet each day I can hardly wait to see what new animals we will find along the trail or in the river. On past trips, we've encountered bull sharks, tapirs, tamanduas, and caiman. On a typically humid September evening in 2017, as we head back to the station, we buzz about the diversity of ground birds we've encountered. We recall the sharply dressed

Opposite page: Ocelots are a rare trophy to be seen by hikers on the Osa Peninsula.

black-faced antwren strutting across the forest floor like a rail might do in a swamp, a pair of subtle marbled wood quail scratching for dinner among a thick carpet of leaves, stately great curassows, and primitive great tinamous. As we gab, a series of high-pitched screeches gets our attention, and turning toward the strange sound we see three tapirs moving fast through the brush. The tapirs I've seen in the past were always silent and never seemed to be in a hurry, but these animals were acting very differently. With a little research, we learn that tapir males are quite vocal when competing for a female in estrus. Apparently, we interrupted two males vying for the attention of a female this evening.

Each walk in Corcovado holds the promise of seeing something new. I guess that isn't surprising, given the park's location on a small bit of land jutting into the Pacific that contains 2.5 percent of the world's biological diversity, though it covers less than one-one-thousandth of the planet. Sparing the park from agriculture (or any other human action that transforms land cover, for that matter) undoubtedly affords tapirs, big cats, and rare birds a critical refuge. Sparing land may also offer a novel way to contain invasive species, such as kikuyu grass, which is common in pastures but unable to invade mature forests. But if we consider a landscape as a portfolio that must both conserve the biota and nourish the bodies and hearts of the people who live there, then the land spared in Corcovado should be compensated elsewhere by intensively farmed land that provides its owners with a satisfied life.

I see the appeal of linking land sparing with agricultural intensification, but I remain skeptical of its advantage and practicality in a wild place such as the Osa. Sure, shy and rare species find refuge in the reserved land, but the surrounding industrial agriculture makes it difficult for wildlife to travel beyond the refuge (see chapter 7). Moreover, the remote setting and limited infrastructure of a wild place may change the economics that favors corporate investment. To me, there seem to be too few workers, roads that are frequently impassable, and inadequate power and other infrastructures to support economically viable industrial farms. Indeed, other than extensive, low-tech ranchland I find no intensive crops or plantations here. I'm eager to learn more about farming in this far-flung setting and am fortunate to have discovered an experiment in progress that aims to demonstrate how farming and wildlife can coexist.

Osa Conservation is a nonprofit organization that runs a small biological station just ten miles from Corcovado. When I visited their Piro Biological Station in 2016 and 2017, they had recently purchased an old cattle ranch and were busily restoring the old pastures to forest for wildlife, improving the mile-and-a-half-long beach for nesting sea turtles, and creating a small organic farm to feed station staff and provide a platform from which they can explore the role of agriculture in conservation. In part, this conservation group is interested in discovering if sharing farmland with wildlife can provide food for people and habitat for rare species. But creating an organic farm in the wilderness was also a practical decision. Though the road to Puerto Jiménez has been widened and leveled since I first traveled it in 2011, it still requires four-wheel drive at most times of the year. To get back to town for supplies, one must ford the Piro, Carbonera, and Escuela Rivers, which after a typical afternoon rainstorm swell to raging torrents capable of sweeping a jeep off the road. Growing produce will save money, increase food security, and increase our understanding of the feasibility of integrating wildlife into the workings of a small-scale organic farm.

TRAVELING TO PIRO gives me a feel for what conversion of forest to pasture does to the wildlife community. Beyond Corcovado, species that exploit crops, cattle, or open country inhabit the agricultural land. Others thrive on the interface between reserve and farm. Birding pastures add red-breasted blackbirds, thick-billed seed-finches, and slate-colored seedeaters to my list. On occasion, I find something a bit more exotic. In 2016, my class hiked along the southern edge of Corcovado through the Golfo Dulce forest reserve and into a seemingly endless sea of lightly treed ranchland. Grassy savanna bordered the park, affording stunning vistas into expansive mountain forest and bringing my students and me close to the region's most massive vulture.

King vultures measure nearly three feet from the business end of their beak to the tip of their tail. Black feathers on the outer wings and tail form a unique frame around a pure white chest, belly, and legs. Atop this strik-

The striking king vulture is the largest of its kind in Costa Rica.

ing plumage sits an astounding, if grotesque, bare head brightly colored by odd protrusions and wrinkles of mauve, yellow, red, black, and orange skin. Like all vultures and condors, the bald head of the king is thought to have evolved so that these scroungers might more easily clean themselves after plunging headfirst into a rotting carcass. The juxtaposition of undisturbed forest and open country grazed by cattle provides all that this scavenger needs: safety, seclusion, plenty of roost and nest sites, and an occasional feast.

Vultures survey the herds of brown cattle that graze the pastures we hike through. The bovines have cut deep trails into the soft hills to nibble the thick mat of grass down to within an inch of its roots. Cow trails are a fixture of Costa Rican landscapes, where pasture occupies a quarter of the nation's land base. The narrow roads, which we walk, are a mess from the cows' hard hoofs. Vibrant red clay turns to grease and muck that permanently stains our boots and in a few deep spots threatens to suck them off our feet, quicksand-like. This steep, rainy country does not seem like cow country to me, but the cows' heavy udders suggest that my assessment is incorrect. They also indicate that ranching in the wilderness is challenging. There are no homesteads within miles of these herds, and the roads, at least now at the start of the rainy season, are difficult to walk on and impassable to all other transportation except perhaps an experienced horse. I wonder how frequently ranchers meet their charges and if a cattle operation in these hills is profitable. The king vulture is betting that the answers were "rarely" and "no."

Other than the king, we find none of the region's well-known megafauna in the pastures. With few trees, monkeys are absent and perching birds infrequent. We see no trogons, no tapirs, and not a single cat track. As the day comes to an end, we leave the muddy cattle pastures, slosh through the shallow Piro River, and enter the forests and farm at Piro Biological Station.

Howler monkeys welcome us to the station with guttural hoots as the sun slides below the watery horizon. Common pauraques, a relative of the whip-poor-will, carry birdsong into the darkness, picking up after the last cocoa woodcreeper song fades into the tropical night. A delicious farm-to-table meal featuring rice with pork and a vegetable slaw of carrots, cabbage, peppers, and celery matches the magnificent natural concert. The monotony of the ranchland quickly fades from memory.

Each meal at the station uses fare from its farm. Fresh eggs, watermelon, and papaya are breakfast staples. Lunch is a typical Tico meal known as a *casado* (meaning a complete meal, like a married man). It features plantains and fish, chicken, or pork from the farm. Sometimes it is a little strange for visitors and researchers to live and work within their pantry. Brooke Osborne, a Ph.D. student from the United States, for example, studies the links among tree diversity, climate, topography, and nutrient cycling in tropical forests of Piro. She has spent a good part of the past four years at the station and appreciates the farm-raised fare. However, as one who eats little meat, she notes that "it is an odd feeling knowing that the cute pigs you see wandering around the station and give occasional belly rubs are destined for your plate, but of course it's much better than any alternative!" Chips and bread made in the city are among the few processed foods used here. Produce from nearby farms augments what the local field doesn't provide, primarily rice and beans.

The farm at Piro is changing to better provide the needed groceries and become self-sustaining, a necessity of most nonprofit endeavors. In

2016, I met Paola Vargas, a twenty-seven-year-old ecological agrarian who was then building the farm, which was named Finca Osa Verde. Starting a farm is difficult—just getting the infrastructure and daily routines in place take all one's time. But as crops, herds, and flocks multiply, so too does the essential ingredient of organic farming—compost. Paola is obsessed with compost, and rightly so, because without it, she cannot build soil, and without soil, she cannot feed the research staff. Paola piles food scraps, manure from the goats and chickens, and debris from the gardens into a compost heap situated in a covered stall about the size of a two-car garage. The stack cooks under its self-generated heat of decomposition for a few months. Paola and her volunteers spread, turn, and reform the pile every few weeks to make sure each component gets some time in the hottest part of the fetid mass.

Paola and her staff used their compost to build a mandala garden at Finca Osa Verde. *Mandala* is an ancient Sanskrit word referring to the sacred energy of a circle. The circular form of a mandala garden is both functional and philosophical. Beds emanating as rings outward from a central point constitute a mandala garden. Some farmers view the concentric rings of the mandala as enabling life's energy to flow naturally. Certainly, they are beautiful works of land art. But most farmers adopt a mandala arrangement for practical reasons. Paths between the rings easily access the concentric circles of crops, and crops that ripen in unison are typically planted in each wedge of the garden so that they can easily be weeded, fallowed, or otherwise tended. The garden rings are carefully arranged and typically centered on a pond or other water source. Irrigation is easiest in the inner rings, which hold a family's or community's most basic and frequently harvested needs. In Paola's garden, this included beans, hot peppers, ginger, turmeric, and squash. The next set of rings, a bit more distant from the water source, may contain products that a family will sell, barter, or use seasonally. At Finca Osa Verde, this included melons, pineapples, medicinal herbs such as lemongrass and basil, and citronella, a useful insect repellent. Finally, the outermost rings include larger shrubs and trees that require little extra water and buffer the inner circles from storms. Katuk (a shrub also known as sweet leaf that produces edible leaves for salads), chaya (tree spinach), and hibiscus filled this layer of the Piro mandala. Beyond the arcs of the mandala, larger orchards held bananas, sugar cane, papayas, and guava, while a greenhouse tailored soils to the needs of lettuces, sweet peppers, and tomatoes. In total, the gardens and orchards produce 30 to 40 percent of the food for the biological station.

To increase sustainability, the farmers of Finca Osa Verde needed to find a way for their small staff to grow more food and develop a crop suitable for the market. These demands necessitated both a move toward mechanization and research into the best way to nurture a wild vine. The staff still accomplishes much of the work by hand, but the farmers have found that tending rectangular beds and planting large fields enables them to produce more with less. Raised beds built under a plastic roof and watered with a low-pressure drip system are more productive and more easily weeded, composted, and amended than the more aesthetically pleasing circular, exposed beds of the mandala garden. A newly plowed acre is just starting to flush its first crop of weeds, which a tractor will remove in a few hours, a job that would take days to do by hand. The field will then grow melons, squash, peppers, and pineapples, significantly increasing the products available to the biological station kitchen without necessitating an enlarged cadre of field hands.

Straightening the fields and using a tractor doesn't mean that Finca Osa Verde has abandoned its old ways. Committed to keeping an organic farm, the staff are still heavily invested in composting and manual labor. Visitors who want to learn by doing help farmers regularly with tasks that require good old-fashioned hand and back power. Cleaning the goat pens is one such chore that seems always to await me when I visit.

In 2017, the new farm manager, Luis Solís, was all smiles as he offered my class a choice of helping with the baby goats or with weeding and tilling the spinach beds. Of course, the cute goats lured most students. I'd experienced goat duty in the past, so this time I headed for the spinach. My crew spent the morning pulling weeds of all size from the beds, using spades to turn new compost into the soil, amending the rich loam with calcium to buffer the acidity of the manure-based compost, and arranging the drip-watering system. We delighted at a ruddy ground dove nest, complete with a nearly mature chick, that was tucked neatly into the remaining patch of spinach (this variety of chaya grows four or five feet tall and is quite bushy). The rest of the class enjoyed the company of the goats as they spent several hours shoveling goat poop in the tropical heat and humidity. Luis's assistant used the tractor to move load after load of goat shit from the pen to the growing mountain of feces in the compost shed. The manure crew didn't complain too much, but they were quick to hit the showers when we returned to the field station.

Bringing an additional acre of Finca Osa Verde into production will help Luis meet his goal of supplying all the food required by those working at Piro, but with the goal of sustaining the broad ecological missions of Osa Conservation, he is also looking at turning a wild orchid into a cash cow. You probably have the extract from the beans of this orchid, which we call vanilla, in your cupboard. If you do, then you know that the precious dark liquid is expensive. And you may also realize that the price has doubled from 2016 to 2017. Vanilla's high cost—up to $275 per pound of cured beans in 2017—is a result of a terrible cyclone that wiped out about one-third of the crop in Madagascar. Though vanilla is not native to the island, farms there produce about 75 percent of the world's crop. So, when Cyclone Enawo destroyed Madagascar's vanilla farms in March 2017, Osa Conservation partnered with researchers at the National University of Costa Rica to begin experimental cultivation of vanilla on Finca Osa Verde.

Vanilla grows wild in the tropical forests of Costa Rica. Three distinct species climb as vines around the trunks of tropical trees. The host trees provide support, like a ladder, for the flimsy green vines that are smaller in diameter than one's pinky finger. Their reproductive habits are fairly perverse, as far as plants go. The vine grows for several years before flowering, and then each flower opens for only one day. When pollinated, during its day, a flower drops from the plant. But, when fertilized, either from its anthers or as a result of visits by pollen-laden *Melipona* bees, the flower develops into capsules, which we call beans. Vines rarely carry more than a handful of beans, which Luis is quick to point out take nine months of nurturing by the parent, "just like a woman with child."

Growing vanilla takes a great deal of effort. Each vine is planted at the base of a host tree and heavily composted so that the root system can spread

A cultivated vanilla vine begins its ascent of a host tree.

several feet in every direction. Farmers weed the composted area frequently and regularly check the vines for leaf fungus. The *Fusarium* fungus is common in tropical agriculture, and it can kill 50 to 70 percent of a farmer's vanilla plants if left unchecked. Researchers at Finca Osa Verde are currently working to isolate beneficial bacteria and mycorrhizae (fungi that associate with plant roots, aiding their absorption of minerals and water and reducing susceptibility to pathogens) that will limit the abilities of *Fusarium*. They also hybridize native strains with the goal of increasing hardiness and productivity of the vines. If the farm science succeeds, more plants will survive the three years required before flowering. But then farmers must remain ever vigilant so that they find flowers on the day they are open and receptive to being pollinated, which on the farm is done by hand.

Luis was eager for us to see his vanilla plantation. We piled into his well-worn four-wheel-drive pickup and sped up the muddy track from the goat pens to the farm's extensive forest. Latin pop music pumped from the radio, and we bounced along the rough trail, higher and higher, until we came to an orchard fronting a wall of mature forest. At the base of each fruit tree, a single vanilla vine crawled from the compost and plastered its evergreen stem and leaves to the new host. The barely one-year-old vines were only about three feet long, and to me, they seemed vulnerable. But to Luis, they appeared miraculous. As he told us how he looked forward to 2020, when the first harvest will occur, he raised his arms like a prizefighter who just scored a knockout, and shouted, "We will be millionaires!" We all had a good laugh. Though the current plantings may be lucrative (each of the six hundred vines is expected to yield a bit more than a pound of dried beans worth two to three hundred dollars), their profits will not enrich the farmers but instead will sustain significant restoration and research, including discoveries that may enable more Costa Ricans to add the profitable vine to their organic farms.

FINCA OSA VERDE is increasing the sustainable lifestyles desired by the ecologists who come to work at Piro. We all appreciate that our foods are locally sourced and organically raised. But the farm is also providing a welcome corridor and refuge for wildlife. The ten-acre finca occupies less than a tenth of the total land owned by Osa Conservation on the peninsula (five hundred acres at Osa Verde and eight thousand in total on the Osa), and an extensive native plant nursery and restoration research effort is helping to reforest much of the former pasture that the group owns. The mixture of mature forest, developing forest, wetland, beach, farm, and research campus supports nearly all the wildlife one finds within the protected borders of Corcovado.

Manuel Sánchez loves the forests that cloak the land owned by Osa Conservation. His grandfather, who managed the property for cattle and timber production, once owned these woods. When the grandfather sold it to a local conservationist in 1990, Manuel's father retained title to a small part of the original nine hundred acres, including the Sánchez homestead. Manuel has spent his entire life connected to this land. He went to primary school at a one-room schoolhouse on the Río Piro, a short walk from his home. Though he commuted to Puerto Jiménez for high school, he returned to Piro in 2004 as a research scientist with Osa Conservation. At this time the organization was known as Friends of the Osa, and it was this group that acquired the Sánchez ranch from Fundación TUVA, a local conservation group that worked with farmers including Manuel's grandfather to establish sustainable lifestyles that were in harmony with forest conservation. Although he enjoys the wildlife of the forest, it is the turtles that ply nearby beaches that speak to him. Manuel has coordinated the sea turtle program at Osa Conservation for the past thirteen years. He has discovered and helped preserve nests of all four species known from the southern Osa Peninsula here on Piro and adjacent Pejeperro Beach: olive ridley, green, leatherback, and hawksbill.

To walk the beach with Manuel is to step back to a time when conservation was more personal, its tools were simple, and its rewards were heartfelt. We gather after dinner, hike to the beach, and patrol up and down the strand looking for sea turtle tracks. We carry flashlights that emit red light so as not to disorient the ancient reptiles that lumber from the surf and crawl up the beach to dig their nests. We watch mesmerized as a female olive ridley's large back flippers carve a neat hole about twenty inches down through the sand. She then enters a trancelike state and drops ninety-eight round, white eggs into the nest. When she has finished, she covers and camouflages the nest before returning to the sea. Her travels produce two sets of easily recognized, parallel tracks across the beach from sea to nest and back. Because poachers looking for turtle eggs also patrol this beach and could easily follow this turtle's tracks to her nest and steal her investment, we gather the entire clutch and take it back to Manuel's *vivero*. A vivero, or nursery, is a small patch of sand suitable for nesting turtles that are enclosed and guarded against poachers. Poachers seek turtle eggs because their purported power as an aphrodisiac brings a high price. Our simple presence on the beach deters some poaching, as we see when a motorcycle's approaching light quickly turns back after spotting our party, but this is only temporary, so we carry our bucket full of eggs to the vivero.

We bury our treasure in a hole dug by Manuel that mimics the one the mother olive ridley dug less than an hour ago. It sits within a grid of more than two hundred other such nests dug with human hands in the vivero. As

Students release young olive ridley turtles, which set off for a life at sea.

we make our deposit, Manuel calls us over to another nest that is hatching. Baby turtles only a few inches long are popping out of the sand and piling up within the screen cylinder that surrounds the nest. We end the night joyfully releasing the brood and urging the newborns across the wet sand and onward as they struggle to break through the pounding surf and enter the realm of Neptune. If even one survives the perils of life at sea and returns to lay its eggs at Piro, conservation will have been well served.

On another morning I sit with my class and Manuel as we watch a batch of eighty turtles hurry to the surf. The students drift off, and we are alone with the last few slowpokes. We see each of the silver-dollar-sized reptiles enter the tide, even though they are among nearly fifteen thousand Manuel will raise and release this year. If not, Manuel tells me, a black hawk might take them. Only the crashing of waves against black sand breaks our silence. Finally, Manuel whispers "Excellent" and breaks into a broad smile full of pride as the brood now begins life beyond his control. He has done this too many times to count, his entire life it seems, but he meets each event with enthusiasm and deep commitment.

Sea turtles are not the only charismatic megafauna that use the lightly farmed lands held by Osa Conservation. The forests at Piro are home to all five big cats of the Osa: puma, ocelot, jaguarundi, margay, and jaguar. Motion-sensitive cameras placed throughout the area by biologists have photographed each species. We regularly see the territorial scrapes made by puma, the most abundant cat on the peninsula, as we hike the trails around the biological station. We see the critical prey of the cats: peccary and agouti, but we haven't yet glimpsed an elusive feline.

Sometimes, when we encounter the prey typically stalked by cats, they treat us like a predator. A memorable encounter with an agouti showed me the challenge faced by a cat on the prowl. The large, ground-dwelling rodent nosed the deep leaves just off the trail as it sought fruits and seeds that rained down from a feeding troop of spider monkeys high above in the canopy. As I moved to keep up and try for a photograph, the squirrel-like animal would drum the ground with both back feet. This signal was quickly and similarly answered by another agouti, which I could not see. I had the feeling that both kept tabs on my location in this manner. I maneuvered for a clear view of the first agouti when a loud, blood-curdling scream came from its hidden partner. I reflexively turned to see what had caused such an outburst, and when I looked back, the object of my photographic safari had disappeared. A perfect, coordinated alarm and vigilance system duped me and gave this favorite rainforest prey species all it needed to slip away.

Primates, such as the Geoffroy's spider monkeys brachiating high above me, are much easier to observe than the wary agouti. At Piro, we are daily in the company of all four Costa Rican monkey species: mantled howler, Geoffroy's spider, white-faced capuchin, and Central American squirrel. My eyes seem always to follow the youngest monkeys in a troop as they explore the very tips of high branches or ride on their moms. Older babies sit upright on their mothers' backs, but very young ones cling with hand and feet to the underside of the mothers' bellies. Their small prehensile tails curl up from beneath their mothers like inquisitive hands surveying their world.

Populations of howlers and capuchins are widespread and robust throughout their range. Unfortunately, this is not the case for spider and

A Central American squirrel monkey filches a guava from the Piro farm.

squirrel monkeys, both of which occupy small ranges and exist in tenuous numbers. Hunting, habitat loss, and range fragmentation threaten both with extinction. Their use of the farmlands at Piro is therefore especially promising. However, farming in the presence of monkeys can challenge even the most ardent conservationist. At Finca Osa Verde, squirrel monkeys regularly raid the guavas and valuable crops of corn and beans. The farmers tolerate the losses and shoo the wily pests from the fields when caught in the act of thievery. Tolerance for all manner of wildlife allows this farm to augment the mission of nearby Corcovado.

Reptiles and amphibians are also at home on this farm. Red-eyed tree frogs and reticulated glass frogs call from the trees and bushes that line the Piro River, and the puddles formed after an afternoon rain attract mating swarms of masked tree frogs and tungara frogs shortly after dark. Their calls, especially the out-of-this-world whine and chuck of the male túngara frog, add mystery to the night. Snakes often hunt the puddles and bushes for frogs, surprising the amphibians and flashlight-toting biologists. In 2017, our short walk from cabin to dining hall took us within two yards of a huge red-tailed boa constrictor. The snake lay placidly coiled under a dark clump of heliconia resting after a big meal or before a molt. When she stretched out, we estimated her length at twelve feet. She remained unmolested on campus for our entire four-day-long visit. What a neighbor!

Comparing the birds I routinely see in Corcovado to those on the Piro farm further confirms the ecological vitality of this place. Spectacled owls, white-whiskered puffbirds, and black hawks live within the forests of the national park and the farm. Bicolored antbirds, tawny-winged woodcreepers, and gray-headed tanagers pay close attention to the columns of army ants that hunt both farm and reserve. I've only seen the scaly-throated leaf tosser once, living up to its name at Piro, though it likely tosses leaves about the forest floor at Corcovado as well. The importance of Corcovado, however, is revealed in its diversity of birdlife. I've sighted both Baird's and slaty-tailed trogons in the park but not on the farm. Ditto for the rare red-throated caracara, two large woodcreepers (the black-striped and spotted), and the bejeweled rufous-tailed jacamar. The highly endangered black-cheeked ant tanager exists only in the park.

The demeanor of wildlife on the farm and in the park further reveals a subtle but essential similarity in the actions of park stewards and farmers. In both settings, the wildlife is unafraid of scrutiny by people. The big boa ignored our daily traipsing. It is routine to approach ground birds, such as tinamous, wood quail, and curassows, without flushing to cover. One can get within a swarm of army ants to experience the enthusiasm with which the ant-following birds pursue the spiders, roaches, scorpions, and other invertebrates that run with abandon from the phalanx of ants. Doing so is a lifetime experience, replete with a few stings from the ants that we wear like badges of honor.

Ecological relations, such as the facilitation of ant-following birds by army ants or the competition among hummingbirds for a flower's nectar, are familiar sights at Piro. A buzzing low to the ground caught my attention during a short afternoon junket. Suspecting a wasp, I looked for the source and discovered two stripe-throated hermits in an aerial duel. These hermits are small hummingbirds, barely two inches in length, that are often confused with moths because of their drab coloration. One bird revved up and took a passing run at the other from about eighteen inches away, its bill forward, heading at its rival like a knight in a joust. Back and forth the duel went for several minutes until one left with the second in hot pursuit. All of this competition was for exclusive access to a few heliconia blossoms in the scrubby jungle.

Birds of prey, including several species of caracaras, hawks, and forest falcons, are common in and out of the park, and unlike their often-persecuted counterparts on European and North American farms, they never appear suspicious of us, even when we point a long camera lens in their direction. Tolerance of tasty, bothersome, and dangerous wildlife is part of the farmers' core values at Piro. If a farm is a self-portrait of the farmer, then the native predators and pests that inhabit a farm must be the truest reflection of a farmer's soul.

What I surmise of the farmers' souls at Piro is somewhat unusual. Others' intolerance of predators and disregard for the role of key prey in the ecosystem has fueled the decline of white-lipped peccaries, the largest wild pig in Central America and the main prey of jaguars, outside of national parks and private reserves. Carnivore biologist Juan Carlos Cruz Díaz notes that development outside parks and reserves often increases unregulated hunting and poaching of peccaries. The delicious flesh of these small pigs is too tempting. Declines in jaguars follow reductions in peccaries, leading Juan Carlos to conclude: "Wilderness makes a difference to the presence and density of these species."

The regulation of hunting could aid the conservation of peccaries and other declining, tasty animals on and around farms. Great curassows and great tinamous are large ground-dwelling birds mostly found in extensive old forest reserves, such as Corcovado National Park. These burly birds are eaten by jaguars and smaller cats but also are frequent targets of indigenous hunters seeking protein and cultural resources. At Piro, where human hunting is no longer allowed, curassows and tinamous are found in mature forests as well as within younger forests and plantations. Protecting bits of forest from hunters in a tropical landscape may be an effective way for farmers to share their land with iconic predators and their prey.

THE UNIQUE ABILITY of vast reserves to protect shy and sensitive species is the reason that they are essential cornerstones of the conservation movement. The strategy to reserve lands for wildlife and other aspects of biological diversity, though critical to conservation, is to me not sufficient. The place-bound quality of reserves, for example, makes them vulnerable to the vagaries of climate change and invasion by nonnative species, both of which can quickly transform a sanctuary into a deathtrap. Endangered spotted owls attest to this problem, because they were evicted rapidly from protected North American refuges when invasive barred owls moved in. The magnitude of reserves needed to conserve our biological diversity is also immense. Harvard biologist E. O. Wilson argues that humanity should set aside half of Earth to preserve other species. Even if we could agree to such a lofty goal, a recent assessment of the world's ecosystems found that just under half of them still retain enough native habitats to achieve the half-Earth target. Within the ecosystems where the half-Earth goal is attainable, today only 12 percent have reached it.

I am in favor of moving closer to the half-Earth goal, but rather than relying solely on setting land aside for the benefit of biological diversity, it seems to me prudent to simultaneously improve the harmony between human and wildlife use of our world. Wildlife-friendly farms and ranches could extend the effective reach of reserves, increasing their resilience to local changes in climate, invasive species, and, most important, political will to maintain them. An ethic among farmers and ranchers that values native wildlife and aims to sustain the land's ecological well-being, such as I've witnessed at Piro and Mastatal, enhances the ability of reserves to support healthy populations of wide-ranging animals that are naturally rare and frequently travel beyond the borders of even our most substantial reservations. Ecologically minded farmers and ranchers may be the only hope for conservation in the world's ecosystems where half-Earth goals are unattainable, but I believe that these citizens are also essential to the long-term persistence of all reserves, even in those few places where over half of an ecosystem appears safeguarded.

The importance of land tenants to sustainable conservation of biological diversity reinforces my opinion that pursuing a strategy that cleaves the land into reserves to satisfy conservation and intensive farms for the production of food, fuel, and fibers is not the only solution. Proponents of this strategy assume that there are only tradeoffs between food production and biodiversity conservation. But a recent consideration of the human dimensions of conservation and farming suggests otherwise. A team of sustainability scientists, led by Professor Jan Hanspach of Leuphana University in Germany, analyzed more than a hundred farming landscapes of the global South (much of Asia and the Middle East, Central and South America, Mexico, and Africa). Here the pressures to provide food for local people and conserve the rich biological diversity are acute. They asked experts in food security and biodiversity conservation about the social and ecological characteristics of familiar landscapes. Their analysis of the expert opinions identified situations in which food production came at a cost to the conservation of biological diversity, protection precluded food production, and, importantly, where high food production and preservation were both attained.

Professor Hanspach and his team found that conservation of biological diversity was not likely in farming landscapes where infrastructure, such as roads and power grids, was first rate, access to markets was good, and the economic resources needed by farmers and investors was high. The ability to produce food in these settings came at a cost to conservation. I saw this cost in the corn and soybean fields of Nebraska and learned that it is typical of farms and ranches in the global North. In landscapes where the financial holdings of the rich and the poor were drastically different (what researchers term high social inequity) and access to land by local people was difficult because it was owned by a few wealthy or powerful entities, the ability of farms to feed the local population and conserve biological variety suffered. The industrial banana and pineapple plantations that expose poor laborers to dangerous conditions and produce luxury foods for distant markets seem to fit well in this category.

The good news from the Hanspach team was that biological diversity and food production simultaneously improved in landscapes where social equity was high, traditional practices and human health were valued, and farmers had reliable access to local land. Adequate access to land sufficient to support a household was crucial to maintaining social equity and food production. As we have seen in Mastatal and Piro, such small-scale farms can also teem with wildlife. That these biologically diverse agricultural landscapes are located far from urban markets and farmed by self-sufficient, average wage earners is consistent with the Hanspach team's findings. Interestingly, upgrading roads and modernizing farm equipment in rural areas such as Mastatal, which many perceive as a way to reduce poverty, may force a tradeoff between food production and the conservation of biological diversity. Protection also is not served when poor farmers have no choice but to rely only on traditional methods because limited food production often leads to overexploitation of natural resources. The well-being of the farmer is of paramount importance to an agricultural system's ability to both produce food and conserve biological diversity. As the authors summarized, "Avoiding a narrow focus solely on infrastructure development, commercialization, and built capital—and instead also strengthening human capital, social capital, and equity—therefore seems critical for fostering synergies between food security and biodiversity conservation."

THE FARMERS I KNOW in Costa Rica have shown me that earning a living wage while producing healthy foods raised with low-tech and organic methods lead to their well-being as well as to the well-being of the native birds, mammals, reptiles, and amphibians that share their farms. This win-win outcome for people and biodiversity conservation has the side benefit of attracting tourists from afar to come and experience the bounty of the farm—both its food and its wildlife. My students and I have been fortunate to gain such experience. We all treasure the locally sourced meals in the company of genuinely satisfied farmers and splendid biological riches.

At Piro one hardly notices the farm. The newly furrowed field seems insignificant in the face of towering ajo (*Caryocar costaricense*) trees or waves crashing on the adjacent wild beach. The few machines are unable to compete with the din of howler monkeys and scarlet macaws and even the muted barking of house geckos. The morning and evening commutes of hundreds of red-lored parrots dwarf the comings and goings of farmers. The workday's end is signaled not by the chime of the clock but by a collared forest falcon's haunting calls from an already dark woodland. The only lights one notices at night are those flashed by fireflies.

LEAVING THE OSA PENINSULA fills me with the hope that people can live in harmony with land spared from agriculture even where intensive agriculture is rare. Remote landscapes need not be simple checkerboards of reserves and industrial farms. Instead, less intense, small-scale, wildlife-friendly farms that abut reserves can enrich society and nature. Within such diverse landscapes, I found the health of wildlife and people to be enhanced. But the economics that limits intensive agriculture near Corcovado to a few large ranches can quickly change. At this time, multinational corporations may not be interested in expanding oil palm, pineapple, or banana plantations to the Osa because of the peninsula's inaccessibility. I've often cursed the long, rough, and treacherous road from Puerto Jiménez to Carate, but now I see its gravel as a guardian rather than a barrier. Traveling through rural areas may seem inconvenient to those of us who are used to a faster world, but as we decelerate, we can use the time to discover how humans and nature coexist in the slow lane.

NINE

The Luxury of Meat

A WORLD AWAY FROM THE TROPICAL forests and small farms of Central America, cold streams drain high peaks and throw wide meanders through flatter, fertile land. Here in Montana, the high-quality grass, which, depending on elevation, is tender from March through August, proved irresistible to midwestern cattlemen. In the mid-1800s, they began annual cattle drives from Texas and elsewhere to fatten up on Montana's opulent pastures. It wasn't long before huge ranches came to dominate the region, occupying, for example, both shores of the Yellowstone River as it left the national park swinging east to the Missouri. Ranching and farming still dominate the Montana landscape and economy, though the average spread has shriveled a bit. There are more than twenty-eight thousand farms and ranches occupying 63 percent of its vast—147,000-square-mile—landscape. Wheat and other crops, such as those grown in the Three Forks area (see chapter 4), occupy the eastern prairies, but cattle and sheep are kings along the flanks of the scattered mountain ranges that rise from the prairie. Unlike in the Midwest, where cattle are fed corn and other grains in large stockyard operations, in Montana most cattle eat their natural food, grass. Many graze solely on native grasses, following the snowmelt up in elevation each summer to feed on the young shoots of fescue, wheatgrass, bluegrass, and basin wild rye. Others are more confined on pastures, where alfalfa hay supplements the fresh brome and timothy grass.

Unlike a cornfield, ranchland holds excellent promise for wildlife. Natural vegetation—habitat for animals—can be maintained because the ground is neither plowed nor paved. And because the infrastructure is minimal, ranchland can be extensive; nearly 10 percent of Montana today is nonirrigated ranch and farmland. However, on the traditional ranch, the needs of wildlife and livestock are often at odds with one another. Deer and elk compete with cows for grass to the dismay of ranchers. Fences can block or trap big game and reduce the safety of wide-open lands for prai-

Previous page: A sandhill crane surveys its breeding territory next to the Anderson Ranch.

rie grouse. Conversion of native ground to pasture frequently preempts wildlife migration to traditional lowlands during the winter. The shrubs and trees that grow along rivers make perfect nesting thickets for migratory birds, such as the yellow warbler. However, when cattle are turned out to graze, they seek water several times a day, and in the process, they often trample or overgraze the streamside to the detriment of native birds. To reduce herd losses, ranchers helped champion the idea of predator control, which eliminated wolves from most of the United States by the early 1900s and significantly reduced other large carnivores, such as the mountain lion. No wonder foxes and coyotes are shy and often afoot when seen on a ranch. And wolves continue to find few friends among ranchers. With wolves' reintroduction to Yellowstone National Park in 1995, their control has returned as a conservation issue. For example, in 2015, U.S. government hunters killed thirty-two Montana wolves after lobos killed forty-four cattle, twenty-one sheep, and two horses; ranchers legally killed another ten (in total that equaled about 8 percent of the wolves in the state). Even the burp of a cow is problematic. Climate change due to the accumulation of human-made greenhouse gasses in the atmosphere threatens wildlife around the world. A 2014 study suggests that up to 35 percent of the methane — one of the most potent greenhouse gasses we emit — comes from our agricultural activities, notably the gassy byproduct of a cow's digestive system.

Cow burps aside, some Montana ranchers are far from traditional. I've met some of these folks near the border of Yellowstone National Park. Here, Hannibal Anderson contends with grizzlies and wolves. His son, Andrew, daughter Malou, and daughter-in-law, Hilary, are forging new ways to handle cows, graze the range, and coexist with increasing numbers of predators. Listening to these rugged souls who wear big hats and shitkickers has shown me the importance of tolerance and adaptability in today's agriculture. Their ways of producing beef contrast entirely with that of most ranchers, who graze cattle during the summer and either feed them hay or send them to feedlots to eat corn during the winter. Moving away from corn-fed beef is one way we can help prairie birds, as discussed in chapter 3. Paying a fair price for the luxury of meat raised in harmony with

Wolves, restored to Yellowstone National Park, occasionally stray from their preferred food, such as the elk this duo has killed, and take livestock and kill ranch dogs.

wildlife affords those like me, who are not vegetarians, an ecologically sensitive alternative to corn-fed beef.

LYING FACE DOWN, I extended my arm fully into the rocky soil of Anderson Ranch. The stone my wife, Colleen, and I struggled to remove was several inches wider than the posthole we had thus far dug into the pasture. The hole needed to be deeper, so this bit of glacial moraine had to come out. We took turns scraping around the rock's perimeter with a fence spike, but it consumed the good part of an hour before we could lever it out. With the impediment gone, the digging was easy, and soon we were planting an old wooden fence post, our second of the morning. These two pillars, which we wedged tightly into the earth by tamping the soil and rocks we had just removed back into the hole with a heavy, five-foot-long steel bar, would serve to anchor a new gate. Of course, we had only a vague idea of how to make the entrance from the remaining fence posts and wire that we had repurposed for the job. Consulting with Malou Anderson-Ramirez, who was managing the ranch for her father this day, we figured we could either wing it or summon up the collective knowledge of YouTube. We winged it.

Making a gate was surprisingly simple. The key was to lay the three strands of wire out on the ground first. Then we wound one end of each around a stout end post, nailed them securely, and positioned a smaller support strut midway between each end of the gate. The solid ground stabilized the strut, making hammering the wire to it easy. We picked up the assembled gate, attached the end post to one of our newly planted fence posts via two wire loops, one at the top and one at the bottom of the pole, and pulled the other end tightly toward the second planted post. Keeping the fence taut, we wrapped each loose end of the wire around the second post and nailed them in place. We stood back to admire our handiwork.

The gate would make moving the occasional carcass and other food waste around the ranch a bit easier. A straighter path along open ground now replaced the long way through aspen. Grizzly bears, which frequent this pasture, are easier to see in the open than in the trees, and keeping offal away from the herd helps separate the cattle from the bears. But that convenience is only part of the gate's benefit. Now cowboys can spend more time with the herd and control better their movements on the ranch, both of which are key to raising beef in the company of carnivores.

Malou and the rest of the extended Anderson family are pioneering a new relationship with predators. Their ranch, one of only five in the rugged Tom Miner Basin, covers fifteen hundred acres. Just beyond Black, Sheep, and Dome Mountains, a mere five miles away as the bear walks, is the boundary of Yellowstone National Park. An active wolf den is on federal land next to the Anderson property. Both predators are regulars at the ranch. Grizzlies, turning the topsoil in search of wild caraway bulbs, till the pastures. Wolves have redefined the goals of the ranch. The Andersons raise cattle now because the few hundred sheep they formerly raised were irresistible to the nearby wolves. Losses of sheep were unsustainable. Losses of two working dogs—B.J., a great Pyrenees, and Spud, a border collie— were heartbreaking. However, rather than calling in the federal posse to kill the wolves, the Andersons turned to their core values.

These values start with Hannibal Anderson, the patriarch. Hannibal

My border collie, Bella, admires the gate Colleen and I built on Anderson Ranch.

is different from most ranchers I've met. He rarely dons a Stetson (though
he is partial to a smaller brimmed Ecuadorian straw hat), and his boots have
Vibram soles. I met Hannibal and Hilary, his daughter-in-law, three years
earlier, having asked them to talk with my University of Washington field
class about coexisting with carnivores. What I heard blew my mind and that
of my students. Hannibal sees ranching as an intentional win-win relationship between the land and people. He is quick to point out the parallel between his views and those of earlier Native Americans. Everything he does
on the ranch supports the health of the landscape, which includes people.

    I feel as if I'm listening to Aldo Leopold when I hear Hannibal dis-
cuss ranch principles. His ethic stipulates that the land reciprocates one's
actions: take care of the ground, and it will take care of you. In practice, this
means grazing cattle in a way that benefits the soil, the range, and the health
of the people who eat beef. Short-term, intense rotational use of many pastures sustains range health. Cattle are treated humanely using low-stress
handling techniques that allow for lifelong grazing on pasture grass. The resulting lean, hormone- and antibiotic-free beef is healthy for the consumer.

The cattle are also much less vulnerable to wolves and bears because, to graze the range appropriately, cowboys closely watch them. Consumers' demands for ecological and health values drive Hannibal, which turns out to be a good thing for bears and wolves as well.

Hannibal was a baby when his parents moved his four brothers and him from Wisconsin to the Paradise Valley of Montana. Three years later, in 1955, they moved southwest a few miles to Tom Miner. Hannibal spent fifth grade in the one-room stone schoolhouse that still stands on the ranch. However, his formative schooling was done back East, first at a boarding school in New Hampshire and then at Harvard. He was a longtime teacher, then principal in nearby Livingston, Montana, before retiring as the district's superintendent. The blend of Ivy League perspective and Montana practicality is evident in his ranching philosophy.

Hilary has a degree in wildlife science and works closely with wolves. She has had a lifelong love of wild animals, horses, and distance running. Family vacations and the chance to improve as a runner by training at high altitudes initially lured her west from Detroit. Attending college at the University of Montana enabled her to live full time in the mountains, and she has never looked back. Now a busy mother of four, she and her crew of range riders keep in close contact with the ranch's cattle with an eye toward reducing their vulnerability to predators. They also trap, tag, and monitor the wolf packs around the ranch. Hilary excels not only in the technical aspects of observing carnivores and proactively reducing their encounters with livestock but also in the social dimensions of community life that are key to spreading the Anderson land ethic beyond their ranch.

To reduce polarization between ranchers and conservationists, Hilary helps ranching groups, such as the Tom Miner Basin Association, to develop a shared goal for livestock management that also supports conservation. She starts on the common ground; for example, the association agreed to work toward a goal of *reducing livestock losses to predation*. The emphasis here is on the livestock, not the wolf or grizzly. Simply uttering the word *wolf*, Hilary notes, is divisive. As members became comfortable with this goal, they expanded it to: *reduce livestock losses to predation, while maintaining thriving wildlife populations on the landscape.* Not all ranchers in the basin are on board with this evolving attitude; some would prefer to shoot a wolf at first sight. But some stubborn old-timers are moving faster than you might expect.

Colleen and I were happy to have been ranch hands for a day and put our hands in the dirt that is sustained by the Anderson land ethic. We were tired, windblown, and dusty, but we were proud of our gate. We understood a bit more why many farmers and ranchers value seeing their mark on the land. We also appreciated the reward that comes from a day working with your animals. Our three border collies mostly explored the ranch and dug after Richardson's ground squirrels as we worked on the gate, but when we needed to move the herd of goats into the upper pasture, they swung into action. We ran as one, agrarian and canine, pushing the goats upslope, facing down the stubborn old billy, and keeping a watchful eye out for grizzlies.

EIGHT MONTHS AFTER my stint as a ranch hand, I'm again bouncing along the seven miles of gravel road that winds its way along Tom Miner Creek to the Anderson Ranch. It's Friday, May 13, and despite the cold, the day seems anything but unlucky. I am anxious to do a bit of birding on the ranch to see for myself if Hannibal's land ethic is sustaining grassland birds. As I top the hill before Malou's house, I notice that the neighbors have encircled a small herd of cows and calves with red streamers fluttering from a white electric wire.

The turbo fladry I'm staring at is one of the tools that the Tom Miner Basin Association uses to reduce cattle losses to predators. The electric shock the fence delivers keeps a hungry wolf out of the stock. The newly born red calf with white chops and mottled cap that is bedded just inside the fladry is lucky to have a rancher willing to try new ideas. This willingness came slowly. Last year, this ranch did not use fladry, and wolves harried the cattle. Rather than adjust his care of the herd, the owner focused on wolves. As is his legal right in Montana, he shot a wolf that passed through the property. However, because of the success of the association in building community and demonstrating success, this year the rancher called before getting his cows and calves and wanted flags. Hilary never expected

A new calf in Tom Miner Basin stands within the protected confines of turbo fladry.

him to take a progressive approach, but he did, and the wolves and calves have benefitted.

It is cold and spitting rain as I turn in to the Anderson Ranch. The rural landscape puts me at ease. Pastures, lined with white-trunked groves of aspen, march uphill to merge with a distant sea of sage and evergreen forest. Several white-tailed deer are grazing on the smorgasbord of pasture grasses and flowers—white pearly everlasting, pink pretty shooting star, and purple camas. I can hear their hoofs pound the hard turf as they spot me and bolt into the safety of the timber, stiff-legged with tails flagging as if pursued by an ancient predator. The *woosh, woosh, woosh* of a passing raven's wings pulls my attention skyward. I'm racking up birds, but like in the Midwest most I first see are typical of wooded or suburban settings: dark-eyed junco, house finch, hairy woodpecker, American crow, and black-capped chickadee. And then the music begins.

Sandhill cranes are trumpeting the end of the day. Their throaty rattles are one of my favorite natural sounds. As evening approaches, they are leaving the pastures and seeking safe roosts among the ponds and rivers.

This big bird—adults stand five feet tall—needs big, open spaces, such as the rich pastures around me or the grain fields farther north and east.

It is still light at 8:00 p.m. as I set up my tent for the night. I'm sharing camp with a cow and calf, three goats, and a dozen horses. The larger animals are ensconced in corrals, but the smaller and more vulnerable goats' night pen is fully predator-proof, the wooden rails reinforced with electrified mesh. I'm feeling a bit exposed camping outside the fence as I glass the dimming pastures for wolves or bears. A large, very dark grizzly strolls out of the timber several miles away on Black Mountain, heading downhill. This bear appears to be the young boar that has been sharing the Anderson pastures with a sow and cubs over the past few weeks. He's working the high slopes, grazing, before settling behind a fallen tree for a nap. An hour later the coyotes are singing all around me, and then the wolves howl. The land is reciprocating Hannibal's care.

Wilson's snipe are the last birds to join tonight's concert. These stocky shorebirds are packed into the swampy bottom of Grizzly Creek, just west of me. What I'm hearing is male snipe dancing in the ranch sky. Like adrenalin junkies, the birds climb high and then plunge quickly. Their spread tail feathers comb the air and produce a "haunting, tremulous sound," known as a winnow. Tonight the winnows are everywhere, sounding more like an alien invasion than a bird display. I drift off as the snipe party finally winds down.

At 5:00 a.m. the snipe are back at it. I'm guessing that a dozen or more dancing males have awakened me. Their winnows start at a low pitch that rises rapidly as the birds gain airspeed. Each sound ends abruptly as the dancer returns to earth, but the number of performers is so high that the air is full of music with hardly a break. It is cold this morning; the temperature dipped below freezing last night, and low clouds stand heavy over the mountains, resisting a slight breeze. From the warmth of my sleeping bag, I hear the avian world awaken. Soft *churrs* of mountain bluebirds soon join the winnowing snipe. A wall of robin song erupts from the aspen. The mechanical tinkling of a savannah sparrow drifts in from the pasture.

The bird community of Anderson Ranch supports my hypothesis that agriculture done right can promote diversity. This morning, as I bird from the old schoolhouse to the upper pasture gate that Colleen and I installed last summer, I encounter a decent grassland set of birds as well as those typical of mountain woodlands. Savannah sparrows dominate the grasslands, but I find a vesper sparrow and a western meadowlark also singing rights to their pastured territories. Mountain bluebirds pounce on insects among a field of shooting stars. From the aspen grove comes the tap-tap-tapping of red-naped sapsuckers, northern flickers, and downy woodpeckers. Pairs of tree swallows and mountain chickadees are perched near old woodpecker nests, ready to renovate them to their standards. Around the ranch buildings, finches—house, purple, and rosy—work the feeder. By the time I'm done birding I've tallied thirty-seven species, every bit as many as one might find in nearby protected areas. Even more impressive are the eight mammal species I've seen, among them many traditional enemies of the farmer, such as ground squirrels, weasels, bears, wolves, deer, and elk.

In this corner of the Tom Miner Basin, the rhythm of ranch life is in harmony with the abundant bird and mammal communities. The ranch hands tie the daily penning and release of dogs, goats, chickens, horses, and cows strictly to the seasonal flux of pasture and predators. Perhaps there will be more conflict when the herd of a hundred or so coming two-year-old cattle arrive in two weeks for their four-month stay. However, I guess the cattle will work the pasture much like the deer and elk do today, leaving room for the sparrows and larks. My concern for ecological sustainability has waned, but I wonder how such a light touch on the land is *economically* sustainable. Malou's husband, Andres "Dre" Ramirez, provides an answer.

Dre is putting the finishing touches on an old sheepherder's wagon that he has converted into a Tiny House. The Conestoga-like contraption has everything that a couple of adventurous tourists would need: beds, a kitchen, a toilet, and a great view of a pasture that sports wolves and grizzles. This wagon is the last in the fleet of accommodations that Malou and Dre are running as an Airbnb to supplement the ranch income. Visitors come from around the world to stay in the old schoolhouse or one of three Airstream trailers scattered around the ranch. As I slept in my tent, a couple from China occupied one camper and Australians were in the old school. The income generated has allowed Dre to quit his job as a cowboy on a neighboring ranch and work full time at Anderson.

In addition to diversifying the income stream on the ranch, Hannibal and his wife, Julie, have worked off the ranch for most of their professional lives. Julie still works as a family practice doctor, providing advice to rural Alaskan communities above the Arctic Circle by phone and annual visits. As was the case in Nebraska and Costa Rica, farm life rarely is enough to sustain an entire family, but with enterprise, the land near and far provides enough for a wonderful lifestyle.

The herd of cattle heading for the Anderson Ranch is a small part of a more extensive operation involving Hannibal's son, Andrew, and Hilary. This husband-and-wife team—part cowboy and part wildlife biologist—has the drive and know-how to meet the challenges of sustainable ranching on the northern range. They are opening up their summer quarters in the Centennial Valley, so despite the threat of late spring snow, I'm heading their way. It's only fifty-five miles as the crow flies from Tom Miner to the Centennial, but to get there, I have to go through Yellowstone, in and out of Montana, Wyoming, and Idaho, over Red Rock Pass, and travel about an hour on the only sand-and-gravel road through the valley. It will take most of the morning. Good thing I upgraded the rental car!

I BUMPED UP AND over the Continental Divide without difficulty, eased past Nemesis Mountain, and across Hell Roaring Creek into the Alaska Basin of the Centennial Valley just east of Red Rock Lakes National Wildlife Refuge. What a place! A wall of nearly ten-thousand-foot-high peaks—the spine of North America—marks the southern border of the valley. Springs, ponds, and creeks dot the marshy lowlands, while ancient sage-covered dunes roll off to the northern horizon.

Winter was still battling spring in this remote valley, the optimistic green of new grass splashing across the tan, ocher, and purple canvas of last year's decadence. Shimmering ahead on the North Valley Road was the first sign of human occupation I'd come on for hours, an old two-story timber house and corral. Across the road, a dirt two-track disappeared into the sage next to a notch where another old structure was perched. It was just as Hannibal had described the setting of Hilary and Andrew's summer home. I drove into the bustle of early season ranch life at the J Bar L.

Hilary's parents were putting the finishing touches on a renovation that had expanded the century-old house to better accommodate four kids,

The Centennial Range towers over the lands grazed by the J Bar L Ranch in south-western Montana.

a cat, and four dogs that would arrive shortly. They unloaded the first boxes and filled the new flowerbeds with a truckload of fertile soil. Moving is something Hilary and Andrew know well; each summer they live here and manage about two thousand head of cattle. As the grass dries, they head north to overwinter along the Jefferson River in Twin Bridges, Montana. In the Centennial, the four kids, who range in age from under a year to eight and a half, help with ranch chores and enjoy fishing, playing with the other ranch kids (the valley has ten kids this year), and bouncing on the trampoline with their nanny, who is a gymnastics coach. In the small town of Twin Bridges, life is more typically rural.

Hilary arrives just after noon with the kids, cat, old housedog, three horses, and living essentials. After we get everything sorted, we enjoy a hot dinner her mom had packed. Andrew arrives after dark with the chickens, working dogs, and another mountain of boxes. The three border collies know the routine and quickly stake out their places on the new deck. We get the hens into a makeshift coop and save the boxes for tomorrow.

My campsite is a short hike into the sage hills. The thick, old grass makes a comfortable bed, and I sleep soundly until the lightning, thunder, and a mix of rain and sleet awaken me just before dawn. The morning is otherwise quiet, with no birds chorusing to greet the day. As the rain stops, the land comes alive with a high-pitched melody of whistles, buzzes, and trills. The songs of meadowlarks, vesper sparrows, sage thrashers, and Brewer's sparrows lure me from the tent. In a brief break from the rain, the birds perch on fence posts and old sage branches to preen and dry off from the early morning soaking. Birds pack into the grass and sage habitat. I quickly tick off twenty-nine species, including those like the bobolink and greater sage-grouse that indicate a healthy grass and shrub ecosystem. I am surprised to find yellow-rumped warblers and hermit thrush among the typical inhabitants of this open habitat but figure that these early migrants are merely seeking refuge from the storm.

WITH THE HOUSE organized, Hilary and Andrew start working with their crews to prepare for the cattle, which will arrive next week.

Hilary and her lead range rider—a woman who will spend the summer tracking predators to gain the information ranchers will need to anticipate and reduce conflicts with wolves and bears—have placed a radio transmitter on a local wolf and are tracking its movements. Finding the den and most frequented hunting grounds will allow ranchers to stay away from certain dangers. The habits of the wolves are only relevant in so much as they inform ranching. Hilary and her riders focus on how to adjust cattle, not wolves. Two more riders will join their team in the coming month so that all four can work the stock with an eye on the wolves through July. Their presence increases the frequency with which ranchers can check in on each of their herds. Because most predators are fearful of people, frequent and sustained herding of the cattle is a straightforward way to reduce their contact with predators. As situations arise, the riders may encircle small pastures with fladry or reduce the stock's use of dangerous ground. In August grizzlies become more of a concern than wolves, so the range riders are attentive to preventing the loss of cattle to the boggy ground, poisonous larkspur, or ailments, such as pneumonia. Animals that die on the range in late summer quickly attract scavenging bears, which may then become predatory and

A recent death in the Greater Yellowstone Ecosystem quickly attracts the attention of a diverse scavenging community, here including a coyote, a grizzly bear, and an unkindness of ravens.

compound herd losses. The range riders' job is to prevent such deaths by quickly removing natural casualties. Doing so can be dangerous, so before approaching a fatality, the riders may use a drone to sneak a peek and make sure no bears have beaten them to the meal.

Hilary obtained funding for her crew from a consortium of federal, state, and private sources. Interested ranchers can get her services free, though many provide trucks, hay, trailers, ATVs, ideas, and cash to the cause. As in Tom Miner, the local rancher association, here known as the Centennial Valley Association, is an important vehicle to administer Hilary's program. This year, Hilary's team is riding the range for seven ranches; only one family in the association remains uninterested in the program. A fully funded program is vital at the start to ease ranchers away from the old practice of merely turning out their stock in the spring and collecting them in the autumn. Hilary's goal is for ranchers to gradually adopt and support a new way of doing business, one that includes near-daily contact with the herd so that cowboys can anticipate and avoid rather than react to conflicts. She is well aware that such change comes slowly. The system most ranchers use today evolved over the past century, and few are willing to quickly turn loose the culture that their parents and grandparents instilled in them.

In a world that increasingly favors humanely tended cattle raised on natural pasture, the switch from traditional to progressive ranching is affordable. Rather than raising hay for the winter, which occupies the traditional rancher's day, the progressive rancher spends more time on horseback, moving cattle among a series of natural pastures. The progressive also strives to tune the stock to the local ecology, for example, calving in spring rather than winter so that fresh grass rather than stored hay can support mother and calf. These adjustments reduce the need to water, harvest, and put up hay so that traditional ranchers can reinvest time and human resources into herding and range riding. The quality of ranch life and the ecological function of the range increase as ranchers become progressive in their herding. They invest less in weed control, bolster the carbon storage capacity of their land, and reduce the need to control predators, all of which grows their bottom line as it increases the coexistence between stock and wildlife.

Andrew is a testament to the ability of ranchers to change. His schooling and first jobs prepared him to be a traditional rancher. But the Anderson land ethic instilled by Hannibal gave him the open mind needed to embrace the notion of holistic resource management as soon as he heard about it. The idea behind this approach comes from Allan Savory, a Zimbabwean ecologist and farmer who espouses that livestock grazing increases biological diversity when it simulates the way native grazers crop their wild pastures. In the Centennial Valley, the once abundant but nomadic bison grazed the extensive sage and grasslands hard before moving to greener pastures. It might be a year or more before they returned. A progressive rancher willing to move his or her stock in an organized and frequent manner can mimic this sort of intensive grazing followed by sustained rest. But to pull it off requires either extensive fencing or dedicated horsemanship. Andrew and his crew of four cowboys prefer horses to fence pliers. They didn't earn degrees in livestock management to build fences, and riding with their stock allows them to observe how the range and the animals are responding to their actions. Besides, fencing pastures within the twenty-eight thousand acres of land they work is impractical.

Andrew's approach to grazing keeps the range healthy and reduces conflict with predators, but his way of handling cattle is designed to further both causes by also rekindling the herd instinct. The low-stress livestock handling techniques Andrew uses come from the research and teachings of pioneers such as Bud Williams and Temple Grandin. The overarching principle is to encourage the cattle to move by letting them respond on *their terms* to a cowboy. Rather than whoop'n and holler'n while forcing animals considered dumb brutes to move, low-stress cowboys approach the herd from the side, riding a straight, visible, and predictable line that gently pressures the cattle to move. As the herd moves, the pressure is released, rewarding the cows' decision. Cattle learn that the herd is safe and that by moving they can control the pressure exerted by the cowboy. Andrew and his crew are often moving cows and their calves, but before doing so, they always allow the calves to join their moms, what they call "mothering up," to again reinforce the herd instinct and also reduce confusion, which otherwise could increase a calf's vulnerability to predators.

Shaping the behavior of the herd is only one strategy at Andrew's disposal; selective breeding is another. Typical Angus are hybridized with smaller cattle, such as Hereford, Galloway, and Devon, to better fit in with the harsh seasonality of the Centennial. These breeds are smaller in stature but calve in synch with the growth of new spring grasses and produce leaner calves needing less of a mother's milk. As a result, the young cattle can grow efficiently on the native grass that is abundant in the valley and overwinter on the windswept prairies around Harlow, Montana. Grazing entirely on grass is less damaging to the environment—it requires no corn, hay, or protein supplements—and results in a high-quality product that the J Bar L markets as Yellowstone Grassfed Beef.

AS I DROVE INTO the Centennial Valley, I figured that the fences I saw were built to keep cattle out of the national wildlife refuge. As I learned about grazing on the J Bar L, I realized that they were there in part to keep cattle on the refuge. Here responsible ranching and wildlife management go hand in hand. Here, in stark contrast to their endangered status throughout the western United States, grasslands and sage hills add to the economy and provide the resources needed by a complete community of birds and mammals. Species such as the bobolink and meadowlark find ample grass within which to nest. Sage-grouse continue to use traditional open areas for their ancient dance parties. Sage thrashers and Brewer's sparrows find shelter in the shrublands. And through all these habitats roam wolves, grizzlies, and cattle. By working together with governmental managers and private conservation groups, progressive ranchers can enhance the land and turn a profit.

It had been raining most of the day, and I was starting to wonder about getting the rental car back over Red Rock Pass and on to Bozeman. If the dirt turned to mud, I heard that the North Valley Road could get greasy. The rain showed no signs of letting up, so I decided to head out while I still had a chance. About a mile from the J Bar L, two pintail ducks that had landed on the road suggested I might have waited a bit too long. I trudged on with no problem through the dunes. And then I hit Elk Lake Road. The next two and a half miles were a total mud bog. At times my car was floating in what felt like a super-thick milkshake. I'd have zero traction one moment and then slowly regain solid ground. As I worked to keep my momentum up, I'd slide on and off the road, sometimes facing forward, sometimes sideways. I had no idea when the mud would end, but I couldn't stop to reconsider my decision. I plowed ahead, figuring I would soon be walking the now ten miles back to the ranch. I finally found stable ground at Red Rock Pass Road. And when I got back to Bozeman, I celebrated by buying an incredible bone-in prime rib at the local co-op that was from a J Bar L steer. It was juicy but firmer than the typical corn-fed prime rib. The flavor was superb and the fat minimal. I paid a premium for this local, sustainable beef, but to me, the twenty-dollar-a-pound price tag was a small cost for the ecological benefits I saw at the J Bar L. As I ate, I could hear the thrashers, sparrows, and larks singing.

HECK THE OLD BORDER COLLIE slinks across the corral to get the stray calves with a simple command of "Bring it!" It is now two months after my

mud race, and Colleen and I are back in the Centennial Valley to see firsthand how low-stress livestock handling works. It is branding time on the
J Bar L, and 120 calves, 120 mothers, nine cowboys and a cowgirl mounted
on horses, and a few stock dogs, such as Heck, crowd the old wooden corral on the North Valley Road. Two huge, red shorthorn bulls have jumped
the fence around their pasture to join the support crew watching the cows.

In the corral, the cowboys and cowgirl are using head-and-heel roping to coax each calf to the center of the arena. With this procedure, one
roper gets a lasso around the calf's neck and begins to walk it away from its
mother so that a second roper can drop a noose on the ground and catch the
back legs of the calf. Short back-handed throws of the sixty-foot-long rope
allow the riders to quickly dally up—that is, get the rope cinched around
the saddle horn so the calf can be controlled, like a big fish fighting against
a tight drag. The two horsemen pull in opposite directions to hold the calf
securely for Isaac and Chase, two of Andrew's crew, who rush in and turn
the 150-pound calf on its side for a quick inspection, inoculation of anti-
viral nasal spray, and branding. Andrew's five-year-old daughter, Elle, gets
the honor of painting the calf's nose with a pink marker, so that the riders
know to ignore it. As the day wears on, pink noses outnumber clean ones
in the herd.

The bellowing of calves amid the chatter of the cowboys blends with
the smell of burning hair as each calf gets a crisp J-L on its rump. There
are no fires or old-fashioned branding irons in use; the crew applies today's
brands with electric heating elements not unlike those you might use to
start a bed of charcoal for the grill. A portable generator provides the necessary power. The whole operation is fast: both ropes are usually in place
within thirty seconds, and the calf is back on its feet and looking for mom
in under three minutes. This low-stress approach contrasts with the traditional snag and drag branding operation that works with the calves already
isolated from their mothers and uses only a single lasso to snag the calf's
rear legs and quickly jerk it to the ground and drag it to the branding iron.

Despite the dust, heat, and horseflies, Andrew's branding crew is all
business. They wear tight, pressed jeans and western shirts buttoned up to
the collar. The early season sun has already tanned their faces, at least the

Andrew Anderson, left, brands a calf at the J Bar L, while another calf roped around its heels and head awaits its turn.

lower portions where the wide brims of the straw cowboy hats provide no shade.

The ropers are part of a supportive community we find around the corral. A few of the cowboys are from the Two Dot Ranch, north of here. The cowgirl and her husband have driven four hours from the Paradise Valley, just beyond Tom Miner Basin. Jim Head and Urs Schmidlin are here from Horse Prairie Ranch. Urs, the ranch manager, is Swiss. He's been roping and riding in the United States for nineteen years, living the western dream. When I tell Jim I never saw him miss a throw, he insists that his roan horse, Danny, makes him look good. I sense that these guys would prefer to work cattle on horseback than just about anything. Tomorrow the whole bunch, Andrew and his crew included, will drive north to the Two Dot to repay the favor they've received from the team.

Everyone takes a break as several ranch hands and Hilary arrive with a picnic lunch. Kids have faces full of watermelon, while the cowboys refuel on chili, cookies, and cake. Smiles are everywhere.

The road back to Bozeman is dry and firm today, but the smell of singed hair stays in our noses all the way home.

THE COWS AND CALVES left the Centennial Valley in early November. It was a good season. Of the two thousand yearlings, cows, and calves that Andrew, Hilary, and their crews worked, they lost only five, all to natural events such as heart failure, leg injury, and disease. There were a few more early losses on the calving grounds before the herd came to the Centennial. Including these losses, 6.5 percent of the herd's calves were lost, including one calf snatched by a coyote. Once in the Centennial, the young animals increased from about 100 to 480 pounds. The conception rate of cows was 96 percent. Andrew was pleased with both of his herds' accomplishments.

Hilary was less enthusiastic about the success of her actions. The hard work of cowboys and range riders that provided nearly daily contact with each herd likely prevented losses from wolves and grizzlies on the J Bar L. But Hilary is quick to point out that they don't know what saved the herd this year. It could have been luck, an abundance of elk or other preferred prey, or the presence of her range riders. This summer also saw some backtracking in the Centennial Valley Association's commitment to nonlethal deterrents, such as range riding. After the loss of a cow, one association member called on the U.S. Fish and Wildlife Service to kill the grizzly bear likely responsible for the damage. This harsh, knee-jerk response rubbed Hilary and the others in the Centennial Valley Association the wrong way because the rancher did not discuss it beforehand. That the other members questioned the response is a good sign, but coexisting with predators in diverse ranching communities remains a work in progress. I hope it continues to improve next year. There are only about seven hundred grizzlies in the entire Yellowstone ecosystem. Their survival as a species in part depends on the continuing development of trust and open communication among ranchers, such as those in the Centennial Valley Association.

Earnings from the sale of this year's finished cattle—mostly two-year-old steers—supported Andrew and Hilary's family, as well as the J Bar L's general manager, Bryan Ulring and his family, and seven ranch hands. Hannibal, Malou, Dre, and their families at Anderson Ranch were also supported in part by the J Bar L's production. But this success was not easily nor quickly won. The first years of holistic management were full of trial and error. However, the structure of the J Bar L allows Andrew and Bryan to experiment without undue fear of failure. This freedom comes from the vision these men share with the ranch owner. Peggy Dulany bought the J Bar L to help conserve the nature that abounds in the Centennial Valley. She believes that grazing in an ecologically sustainable manner is an avenue to producing healthy food for people and a safe environment for rare animals, such as the wolf, grizzly, sage-grouse, and ferruginous hawk. Turning a profit in the short-term was not Peggy's primary goal. But, as is often the case when good people are motivated to do their best for a cause they truly believe in, success is not far away. Andrew had to cull about 20 percent of his herd when he first began to graze the Centennial in 2009. That is twice the rate expected in a traditional operation. But in the last two years, the herd and the land have come together to turn a profit for the ranch.

Hilary calls this style of ranching "living in communion with the land." She sees little value in dominating nature only to serve people. Instead, she sees every part of the system for its inherent virtue, not merely as a resource to be used and managed for people. People are part of the web of life, connected and dependent on all parts of the ecosystem. Only through sharing can the ecosystem be sustained. In the Centennial Valley, this becomes explicit. Holistic ranchers live in communion with all parts of the ecosystem, including the grass, sage, sparrows, tiger beetles, wolves, cows, and people. In contrast, a rancher still holding dominion over nature shares with the ecosystem only pieces that directly and economically serve the rancher—grass, cows, and people. Hilary finds the motivation to dedicate her life to sustaining a ranch ecosystem because she encounters many who appreciate her holistic land ethic but few who seem able or willing to implement it.

As the J Bar L ethic spreads, it is helping the ranchers of the Centennial Valley to succeed in meeting the triple bottom line. I saw evidence of social sustainability in the community that gathered to brand cattle, in the cooperation of governmental, nongovernmental, and private landowners that comanage a national treasure, and in the strong family ties that allow a young family of six to live happily in an austere land. Ecological sustainability oozed from all pores of the valley; I experienced it in the guise of a river otter darting into the sage, the howling of wolves and coyotes, the track of a grizzly, the shadow of a prairie falcon on the wing, the music of thrashers and larks, and the many breeding display grounds, or leks, of greater sage-grouse. I tasted the economic sustainability of progressive ranching in the grass-finished steak I was willing to pay a premium for and saw its future in a calf that caught my attention during branding. Her left ear bore a yellow tag numbered 4006. Her white face and brown eye patches — a mix of Angus and Hereford genes — gave her the look of a masked fighter. She certainly gave the branding crew all they could handle. She seemed to me to have a winning attitude, which might explain how she put on nearly four hundred pounds and her mother was able to conceive another calf during their short stay in the Centennial. The hybrid spirit of this young heifer thrived in the challenging environment of the Centennial, much as I suspect the young and progressive ranchers I met there will also succeed in an environment that increasingly values sustainably produced food.

TEN

Cows as Tools

TROUT POCKMARKED THE SURFACE OF the narrow, gin-clear stream as they slowly rose to slurp insects. Regular flashes of silver below the surface showed that other trout were working the stream bottom as well, gobbling up nymphs of aquatic insects, such as caddisflies and mayflies. The sight of healthy trout finning clean, cold water was perplexing, because this creek drained a recently grazed cow pasture. Cattle are known to reduce a stream's ability to support fish by destabilizing its banks, which erode and muck up the water's visibility with silt; eating or trampling plants that provide cooling shade; and aiding harmful algae and bacteria that thrive on the bovine crap that inevitably ends up in the waterway. As the actions of cattle widen a stream, the water becomes shallow and flows slacken, which exacerbate warming and reduce the current's ability to carve deep-water refuges under the bank that trout favor. Cattle are often fenced from riparian areas to maintain and restore stream health, but where I stood there were no such barriers. Cows were free to drink from the cool stream and graze the tall grasses that steadied its banks. And trout seemed oblivious—that is, until I cast a grasshopper mimic into the lively waters.

As soon as I entered the stream to fish, the frisky trout retreated to their safe, deep waters. These fish were cagy, and they had a refuge. I stepped back from the water's edge and exchanged my hopper for a potentially less intimidating miniature black fly. The response to my second cast was a bit more enthusiastic. Trout quickly rose to inspect my offering, but still, they didn't strike. I flicked the fly into the stream again and again without fooling a single fish. An hour or so later I'd finally hook and release trout more than sixteen inches long, but it took dropping a nymph imitation deep to the rocky bottom to have any success. These animals had seen a lot of artificial offerings, residing as they do on the MZ Bar Ranch. The ranch, just outside Belgrade, Montana, is managed by Tom and Mary Kay Milesnick to support both a trophy trout-fishing business and more normal cattle ranch operations.

Opposite page: Fed by springs, Ben Hart Creek winds through the MZ Bar Ranch.

In the past, the Milesnicks were quick to provide public access to the streams on their ranch, but after the movie *A River Runs through It* glorified fly-fishing for trout in Montana, fishers overran them. To provide some control over access to their land, they incorporated Milesnick Recreation and began limiting the number of anglers per day and charging each a fee. As the business grew, it provided about 6 percent of the ranch income, a decent counterbalance to the reliance on cattle, grain, and hay. In providing for recreational fishing, Tom did not fence cows out of the small river and three spring-fed "cricks" that braid the twelve-hundred-acre property. Instead, Tom adapted his grazing strategies so that he could use his cows as tools to maintain the health of the riparian zones that border them.

That's right, tools. Instead of spraying, burning, or mowing, Tom used precision grazing to keep unwanted plants in check without damaging the soil, fouling the hydrology, or denuding the bank. And this unconventional approach in ecological agriculture was accomplished by a most conventional cattleman. Tom Milesnick, however, has surprised me for years. He enjoys lecturing to my wildlife science classes when we visit the region and never shies away from answering probing questions from my liberal West Coast students. He is equally at home watching eagles or wondering about the deer and pheasants that roam the ranch as he is at making a "jacket" from a dead calf's skin. (Tom places the jacket on a different calf in the hopes of enticing the bereaved mother to adopt it as her own.) Tom wears a huge Stetson, and his bookshelf belies a conservative character. To him, coyotes and badgers are buggers that do not belong on the ranch. Bison and wolves do not interest him. But throughout more than four decades of ranching, his innovations have not only built a vibrant herd but have taken great care of the land.

Tom and Mary Kay might not seem the type to embrace science to guide their ranching practices. But contrary to persistent stereotypes of ranchers, I have found the Milesnicks to be anything but close-minded. Their willingness to try new ideas and adapt their practices as guided by experience stems from a close working association that they maintain with the agricultural college at nearby Montana State University (MSU). There they support and learn about new approaches to ranching, test these on their land, and invite students and professors to see how their ideas pan out in the real world. Their ability to demonstrate how ranching and ecology can coexist earned the couple MSU's Outstanding Agricultural Leader award in 2013. Their integration of rangeland science and ranching has a long history. Tom's dad, Stan, was the first graduate from MSU's range management program. Finishing up in 1942, he was shipped off to Texas for basic military training to aid the war effort. His superiors did not quite know what to do with his expertise, so they put him in charge of the only range they had . . . the rifle range! Tom and Mary Kay also graduated from MSU. Tom took the forty-year-plan, as he calls it, starting in 1966 and graduating in 2006. Mary Kay laughs at this and adds that during this time "he worked hard!"

Stan Milesnick founded the MZ Bar in 1936, when the filling of Fort Peck Reservoir forced him off the land his parents had homesteaded. Tom joined his dad on the ranch in 1972, where they raised Herefords and Simmental before settling on Red Angus. Tom jokes that in this country where nearly every cow now is a Black Angus, he raises reds so that he can tell them from the neighbors' herds. His primary focus is on providing grass and hay to the herd so that at the end of summer he can sell the calves his heifers birthed to midwestern buyers who fatten them at feedlots. As his reputation grew, he started holding back some of these feeder calves and breeding them overwinter so that they could be sold for a higher price the following spring as bred heifers. Tom doesn't finish his cattle on grass as Andrew and Hilary Anderson do just a few miles away in the Centennial Valley (chapter 9). He isn't a big fan of grass-finished beef, even though keeping cattle off of midwestern corn reduces their greenhouse gas emissions, quipping, "It tastes different." But he manages his land with the same short-term, intensive, rotational grazing system that the Andersons employ. This science-based approach derives in part from considering how grasses coevolved with herds of wild grazers, such as bison and wildebeest. And this is part of why his ranching operation coexists with Mary Kay's fishing business.

Cattle are attracted to streams for water, food, and shade. And so, Tom carefully manages his herd's access to these resources. Tom uses fencing to partition his ranch into paddocks so that grazing pressures within

Rotational grazing is timed so that just enough forage is left for the grass to regrow quickly. Here on the left of the fence the grass has not been grazed for thirty days, while on the right the cattle have just been removed from the paddock after three days of grazing.

each can be precisely regulated. A typical paddock is around forty acres, and a few hundred cows will graze it for two or three days—just long enough to reduce the knee-high Garrison creeping foxtail (a vigorous hybrid of quackgrass and timothy) to a height of about six inches without stunting its growth. The remaining grass blade is sufficient to allow the plant to photosynthesize and quickly regain its lost stature. In about thirty days it will again be knee-high and ready for a second grazing. By removing cattle from his pastures before they overgraze their forage, Tom effectively limits the amounts of silt and excess nutrients from entering the stream.

When grazing a paddock containing a reach of a stream, Tom packs it with cattle—up to two hundred head in fifteen acres—but only allows the herd to remain there overnight. The animals are in by 6:00 p.m. and out again by 6:00 a.m., so the "fishermen never even know they were there." Cattle eat the most nutritious forage first, which along Tom's streams is

Gravels indicate a hard crossing that Tom has engineered into the creek bank to allow cattle to drink without silting the water and degrading the stream edge.

either the foxtail he nurtures, the invasive reed canarygrass that plagues waterways throughout the continent (chapter 6), or other weeds such as Canada thistle and houndstongue. By flash grazing the riparian paddocks, the cattle quickly nip off these grasses and weeds, enabling the native streamside sedges (which are less palatable because of the silica that they contain) and willows to thrive. Cattle could trample the stream bank and muddy the water, even in the short time that they are allowed to graze. But Tom has anticipated this possibility and has constructed special crossing and watering areas for the herd. Next to natural gravel bars, Tom has dug down into the streambed, installed highway underlayment, and covered this fabric with gravel to provide his cows with eight- to ten-foot-broad access points. Using these hard crossings, the cattle can move quickly in and out of the water and keep their feet clean. Tom's cows "don't like to get in the mud," so this simple innovation serves the herd and the stream well.

The rotational grazing system seems to be working nicely. While Tom and I survey the stream, an osprey crashes into the water and emerges with a fat trout. As we travel throughout the ranch, I hear sandhill cranes and spot American wigeon, northern shovelers, mallards, killdeer, eastern kingbirds, red-winged blackbirds, a northern harrier, a great blue heron, and a nesting pair of red-tailed hawks. Hordes of swallows—northern rough-winged, tree, and barn—dip low over the stream. Tom whips out his phone to show me the sandhill crane chicks that thrilled him just two weeks earlier, as well as the flocks of snow geese and white pelicans that stopped to forage last winter and early spring. Amid the hustle and bustle of birds, Tom focuses on the stream. He shows me a backwater channel that he created years ago to provide a refuge where young trout can fatten without attracting the attention of their larger and carnivorous brethren. As he talks, Tom is distracted by an unusually large trout noisily gulping insects from the surface just a few feet away in midchannel. This beautiful music—a mix of flowing water, feeding trout, and calling birds—is a fitting tribute to the life's work of an ecologically minded rancher. By now you might be thinking that Tom is an avid fisherman. Not true. His stewardship of the East Fork of the Gallatin River, Ben Hart Creek, Thompson Creek, and Smith Creek is, in his words, "simply the thing to do to protect the resource." "It's all part of the system," he adds.

Tom and Mary Kay are not alone in using cattle as tools to enhance an aquatic system. Throughout Oregon and California, human actions have eliminated vernal pool ecosystems at an alarming rate. Primary causes of the loss are habitat conversion, invasion by nonnative grasses, and, as the Federal Recovery Plan states, "exclusion of grazing in areas where grazing has been a historic land use and inappropriate grazing regimes." Sara Sweet, restoration ecologist at the Nature Conservancy, works with ranchers in the Sacramento Valley to graze vernal pools in support of the plants and animals that require them. The conservancy leases grazing rights to about twelve hundred acres in the Cosumnes River Preserve so that cows can help pack the soil and remove invasive wild barley, which smothers out rare native plants. As we surveyed one pool, Sara pointed out how native sedges along its edge were spared by the cattle, just as they are along Tom Milesnick's creeks. Maintaining grasslands and conserving vernal pools is accomplished by regulating where, when, and how much grazing can occur.

Canada geese, white-faced ibis, and killdeer gather around a vernal pool in California, maintained by rotational grazing.

Vernal pools form in slight depressions in hardpan soil. They are ephemeral—drying up late in the spring or early summer—but are critical for many unique animals and plants during winter and spring. Extinction threatens thirty native plant species and five animal species due to the loss of vernal pools. The threatened animals include unique fairy and tadpole shrimps that complete their entire life cycles within the pools and a ground beetle. Migratory shorebirds and waterfowl, including white-faced ibises, black-necked stilts, and American avocets, flock to the pools to fatten up on shrimp and other invertebrates. When appropriate numbers of cattle congregate in and around the pools, they remove the nonnative grasses that suffocate the native flowering plants and compact the soil, which enables more water to accumulate and last longer into the summer. In so doing, they allow larger animals, such as tiger salamanders, to complete their life cycles here as well.

The more I read, the more I learned about grazing as an implement in the conservation toolkit. In Alberta, Canada, fencing cattle from streams increased the health of riparian vegetation, soil, and water flow. However, the research scientists who studied this system also found that "serious management issues arose" with the exclusion of grazing. These problems included increased fire risk, increased weeds, invasive species, and non-utilization of forage. They recommended periodic grazing to reduce these issues. As we learned in chapter 9, wildlife managers in the Centennial Valley addressed these same problems by permitting grazing in the national wildlife refuge by the Andersons' cattle. Likewise, in Oregon's Blue Mountains, conventional approaches to limiting grazing adjacent to streams bearing endangered salmon, such as fencing, seasonal closures, and provision of upland water sources and salt blocks, while useful, often led to conflict between ranchers and conservationists. The resulting lawsuits helped neither fish nor farmer. There, federal regulators are considering innovations, such as short-duration, high-intensity grazing, that protect fish and provide an economic return to ranchers. Moderate grazing of salt marshes along the Wadden Sea coast in the Netherlands also enhances the abundance of most breeding birds. There, grazing limits tall vegetation from dominating the marsh and reducing its suitability to songbirds and waders, such as the pied avocet, meadow pipit, and Eurasian skylark.

In exceptional cases, even a grain field can be of benefit to fish. Knaggs Ranch, a 2,500-acre spread in California's Central Valley, opens its flooded rice fields to king salmon each winter. By doing so, the ranch provides important floodplain habitat to this fish, which the Endangered Species Act classifies as threatened. Once abundant throughout the Sacramento River basin, chinook—or king—salmon populations have declined as development and agriculture reduced their rearing habitat. Salmon that return to the Sacramento River each winter are among eight species recently identified by the National Marine Fisheries Service as most at risk of extinction. Farmers and researchers are collaborating at Knaggs Ranch to help this species by rearing young salmon (fry) in flooded fields of rice stubble each winter. In this rich rice soup, initial experiments demonstrate that the fry triple in weight in just six weeks. Draining the fields before planting each spring enables the young salmon to swim back into the main channels of the Sacramento River and out to the Pacific. Researchers suspect that these youngsters will survive the perils of this outmigration better than those reared only in the less productive river, but it will take time to tell.

AS TOM AND I MOVE away from the stream toward larger upland paddocks, a sweet song attracts my attention. I quickly spot the singer, a small black bird with a yellow head and white striping on its back that I recognize as a bobolink. This grassland specialist is fittingly called the skunk blackbird in reference to its stripes of white, which appear to be discrete on its black backside when its wings are at rest. Others think that bobolinks "look like they are wearing a tuxedo backward." And their song, often given in flight like that of the horned larks we met in chapter 4, is described as a "bubbling delirium of ecstatic music." That might be a bit over the top, but it is interesting that each male bobolink has two song types and that he mixes phrases from each into long, rambling monologues, such as the one that caught my ear.

As I returned to Tom's fields over the coming months, I kept a close eye on his bobolinks. Late in July, I discovered a few males and what appeared to be fledglings in an area reserved from grazing or mowing. Here the grass I pushed through topped my head. Deer moved to my side and ahead, but the thick Garrison creeping foxtail entirely engulfed the big mammals. My only glimpse of bobolinks was fleeting. The next day, I fared better. Just as the sun was peeking above the Bridger Range, I spotted a male and two fledglings eating wheat that was about a month away from harvest. Around a bend, in the road, another male and a single youngster were also eating the ripening wheat kernels. Finding only males in the company of youngsters was a bit surprising, because bobolinks are polygynous. That is, each male may breed with one or more females, so I'd expect plenty of mothers in association with their young. At this time of the year, however, females are well into their postbreeding molt and perhaps away from their breeding grounds or less attentive to their grown offspring. Tom wasn't put off by the birds' filching of his grain; red-winged blackbirds, savannah sparrows, and bobolinks all took a bit of his profit, but they too are part of the system. Tom's diverse use of the land—a combination of rotational grazing, haying, planting cereal crops, and letting some difficult land quietly lie

A male (left) and recently fledged bobolink in wheat on the MZ Bar.

fallow—allows bobolinks to find safe, productive nesting areas. This is not the case elsewhere.

Bobolinks occur widely throughout Montana, but they are generally rare. And their numbers here and throughout North America are declining. Annual trends from the Breeding Bird Survey suggest a decline of 0.85 percent annually in Montana from 1968 to 2015, and a drop of over 2 percent each year during this time across North America. Researchers in Illinois fear that they may be extinct in their state within the next half century. A variety of factors on the breeding and wintering grounds are responsible for this species's decline. In the United States, federal agents haze and occasionally kill bobolinks when they overstay their welcome in agricultural settings. Farmers shoot bobolinks on their wintering grounds in South America, where massive flocks can eat a fair share of the harvest. People eat bobolinks in Jamaica. But most significant to their plight is the decline in meadows and hayfields where they spend the majority of their lives. In the United States, bobolinks formerly occupied tall- and mixed-grass prairies, but as settlers converted these habitats to agriculture, the birds moved into hayfields and grazed meadows. Their ability to adapt to agriculture is legion, but as grazing and haying have increased in intensity, the numbers of bobolinks that successfully breed have plummeted.

Hayfields of New England and the north-central United States are especially problematic for bobolinks. When the adults return from their faraway winter quarters—migrating some fourteen thousand miles from the pampas grasslands and grain fields of South America—the northern alfalfa fields and grasslands of the United States appear the perfect habitat for nesting. With moderate grazing, these lands offer a rich mix of nesting cover and plants that produce the seeds and harbor the insects that breeding bobolinks require. The adults court, pair, and initiate their nests as the hay grows. But before the nestlings can leave the nest the hay is ready for cutting, and as the swather levels the crop and forms it into windrows, many eggs and chicks—up to 95 percent of those studied in Vermont, for example—are destroyed. Bobolinks fare better if they select fields that ranchers graze rather than mow, but even here nest loss is substantial. In Wisconsin, bobolink density was higher on rotationally grazed lands than on those either continually grazed or spared entirely from grazing. But the productivity of bobolinks and other grassland birds was highest on ungrazed paddocks. In Vermont, where rotational grazing such as that championed by the Andersons and Milesnicks doubled in occurrence from 1997 to 2006, bobolinks that nest among the dairy herds often have their eggs or young trampled or eaten by cows.

Bobolinks can coexist with hay farmers and ranchers, but this requires some sacrifice by the landowner. Delaying mowing, enlarging paddocks, setting aside some land from harvest, and timing grazing to reduce disturbance to nesting birds are all proven strategies. Birds, dairy farmers, and ranchers could all thrive. In Wisconsin, for example, this would occur if landowners managed about two-thirds of their land as intensively grazed small paddocks (approximately twelve acres worked by two hundred to three hundred cows for a day or two every two weeks) and compiled the remaining third of their acreage into an ungrazed reserve. In Vermont, when cutting of hayfields was advanced and then delayed, so that the first cut occurred in early May (rather than early June) and the second cut did not happen until the third week of June, bobolinks and other grassland birds such as savannah sparrows had ample time to nest and raise their young to independence. Female bobolinks that were afforded the time to breed between hay harvests produced 2.8 fledglings each year. This production is quite a boost relative to those that nested where farmers mowed hay every thirty-five days beginning in early June. In the latter typical situation, female bobolinks eked out only an average of 0.05 fledglings per year. Grazing paddocks intensively before the third week of June and again forty-two to fifty days later after most nesting has occurred produced bobolink success nearly as high as did delayed mowing (2.3 young per female).

The advantages of delaying mowing and grazing to bobolinks and other grassland birds in New England have prompted a novel conservation approach. Audubon societies in Massachusetts, Connecticut, and Vermont started the Bobolink Project in 2007. This partnership generates money by crowdfunding and uses it in a reverse auction to pay farmers to delay mowing. Partners solicit bids from farmers, indicating what they would accept to delay mowing. In 2017, farmers were paid up to sixty dollars for each acre where they delayed harvest. In total that year, seventeen farms in Vermont, Massachusetts, New Hampshire, and New York enrolled in the program (lack of funds excluded many more). The delay of 630 acres from mowing produced 820 young bobolinks! Such innovative programs are sure to help bobolinks, enable farmers to live more sustainably on their lands, and engage the public. We can all help Project Bobolink by joining the crowd that funds the study (www.bobolinkproject.com). Or find a similar project closer to home, such as one in California that pays farmers who benefit the declining tricolored blackbirds by delaying the harvest of triticale, a wheat-rye hybrid grown mostly for silage.

Tricolored blackbirds nest in huge colonies—up to two hundred thousand nests jammed into fifty acres—often in triticale fields. The intensively farmed Central Valley of California is home to nearly the entire global population. Thousands of nests are destroyed each year during the triticale harvest, which peaks just before the young blackbirds leave the nest. The once abundant bird that numbered in the millions when it was a mainstay in the diet of pioneering miners was reduced to 145,000 at last count in 2017.

And the 2017 count was a positive sign, reflecting a greater than 20 percent increase over the previous year's tally. This rebound was in part because, for the first time, the triticale harvest took no blackbird colonies. All farmers whose crops were nested within by blackbirds were paid to delay their first harvest, just as were those New England farmers whose hayfields hosted bobolinks. Subsidies such as these benefit farmer and bird alike while providing an interested public the opportunity to help conserve a rural way of life and birds that have come to rely on agricultural lands.

I RETURN TO THE MZ Bar during the summer of 2018, but I hardly recognize the place. New buildings have replaced the old barn, and a flurry of workers tend to new landscaping and construction projects. Typical black cattle now roam where once there were only Red Angus. The grassy crick edges are ungrazed, though still graced with the machinelike sounds of marsh wrens. The other natural wonders of the ranch seem mostly intact—the wren song blends with the *witchity-witchity-witchity* of abundant common yellowthroats and the frequent splash of trout. Kingbirds, both western and eastern, hawk insects from fence lines and dead branches, turned a rich pewter color by the cold winds and bright mountain sun. Waterfowl broods swim in the slow backwater as killdeer, snipe, herons, and spotted sandpipers work the shore. Sandhill cranes stalk the verdant fields of alfalfa. American black-billed magpies command lookouts atop the huge round bales of hay, occasionally attracting the attention of the resident peregrine. Beneath this natural beauty hides a harsh reality.

Tom and Mary Kay lost the MZ Bar four years ago. The ranch that Tom managed with his father beginning in 1972 that included twelve hundred owned and five hundred leased acres along the spring cricks and five thousand high-elevation acres above nearby Livingston, Montana, was sold in late 2013. Tom and Mary Kay dissolved their fishing business and consolidated their farm activities on the hundred acres of the ranch that they owned outright and on which they lived. As I sit in their living room four years later, the sting of their loss is still evident.

When Stan Milesnick, Tom's dad, became ill in the 1980s, he incorporated the ranch to avoid estate taxes on his death. The corporation ran the ranch by consensus of its shareholders, which consisted of Tom and his siblings (three sisters and a brother, none of whom were interested in ranching). Although Tom managed the ranch, the corporation approved his annual budget and controlled the fate of the land. When Stan died in 2013, Tom's siblings voted to sell the ranch. Tom tried to buy it, but he could only afford to pay the agricultural value of the land. The amenity value was considerably higher, so his siblings rejected his offer and sold the ranch for top dollar to a venture capitalist from the San Francisco Bay Area. I can't imagine the hurt that Tom and Mary Kay suffered. Tom admits, "It's really hard on me." Mary Kay adds that they have had "absolutely no interaction since the sale with Tom's mother or siblings." The Milesnicks lost more than the ranch: they lost a considerable part of their heritage. Such loss is increasingly common as the descendants of farmers and ranchers either lose interest in the land or cannot resist the windfall of cashing out as values rise in response to the growth of nearby towns or the desires of wealthy people to own their piece of paradise.

Tom isn't completely divorced from his former ranch. The new owner of what is now called the Brown Rainbow Rooster Ranch considered Tom's advice and hired Tom's son, John Milesnick, to manage ranch operations three years ago. John is an easy-going exmarine who did a tour in Iraq in 2005 and knows the land as only a boy who grew up on it could. After graduating from MSU with a degree in range management and business, the third generation of Milesnicks to do so, he gained relevant experience by working at a nearby seed and grain processing and packaging plant. As I head into the field with Tom and John, the abandonment of Tom's innovative grazing schemes allows me to see better how they worked. The new owner is an avid fisherman to whom the cricks are sacred water. The cattle John manages are no longer allowed into the cricks to drink or graze as they did under Tom's watch. The result is lush vegetation, but it is composed mostly of reed canarygrass instead of native sedges. That will be a substantial future problem on the Brown Rainbow Rooster. Tom laments that his experience "from a lifetime of getting the system figured out" is mostly ignored. John is frustrated by the loss of cattle as a tool to manage the creekside vegetation.

I learn more as we pull on our rubber boots and visit a new off-stream watering site for the roughly three hundred head of cattle John runs on the ranch. He grazes them in the upland pastures just as Tom did before, rotating the herd between nineteen paddocks that are grazed for two to three days and then rested for three to four weeks. The grass here is healthy, but the mud around the watering site nearly overwhelms my calf-high boots. Providing water away from the stream is a traditional approach to keep cattle out of sensitive waters. By concentrating the herd, however, it damages the surrounding land. And the infrastructure required (a short well, solar panels to power a pump, a tank with a ramp for calves to safely use) is an unnatural presence on the property. In contrast, Tom's innovative hard crossings mimicked natural gravel bars on the creek.

Tom and John survey an ungrazed border to Thompson Creek as the sun breaks through a midafternoon shower. Tom predicts this could soon be a mess of weeds. Not only is reed canarygrass getting a foothold here, but soon there will be noxious Canada thistle and houndstongue. These weeds outcompete grass when thatch builds up due to lack of grazing. Thatch smothers the ground much as your grass clippings do if you pile them up in a compost heap. Cattle (and native grazers) keep grass healthy by clipping off some biomass and using their hooves to grind accumulating thatch and waste into the soil, improving its fertility and organic content. Tom would have already grazed this area twice by now in years past, much as bison might have before his grandfather bought the place. John notes that although overgrazing inevitably leads to increases in weeds, "sometimes no grazing can be just as bad." He is proud of the 13 percent organic content of his grazed soil and doesn't want to see that drop. But there are other consequences. Mowing the many acres of streamside grass that snake throughout the property isn't practical, so when the weeds arrive John will have to use herbicides, not cattle, to suppress them.

John is building the cattle herd and growing hay mostly for sale to ranchers and horse owners in the region. Today most cattle are just summer residents on the ranch. Only the dozen John owns will be bred. This change also has unintended ecological consequences. A colleague who studies eagles first introduced me to Tom as we tried to trap and tag the iconic avians on the MZ Bar. Migrating and wintering bald and golden eagles stop by the ranch in large numbers, aiding Tom's efforts in pest control by feeding in early spring on emerging Richardson's ground squirrels and cleaning up afterbirth during the calving season. Though bald eagles have increased dramatically since the United States banned DDT in 1972, the population of golden eagles has fared less well. The proliferation of turbines to generate power from wind, persecution, and loss of prime foraging grounds in the western United States challenge golden eagles. Maintaining the staples these mighty hunters require—ground squirrels, jackrabbits, and carrion—is critical to their continued survival. The large cow-calf and bred-heifer operations that Tom ran at the MZ Bar provided a bounty for eagles that is now greatly diminished. As Tom puts it, "There's something wrong with this picture, and I didn't draw it."

TOM MILESNICK IS A quintessential cattleman. He eats beef "at least once a day." He supports the subsidization of midwestern farmers to grow corn for cattle and ethanol, the waste products of which also feed cattle. Just before he lost the MZ Bar, he was working out the logistics of bringing back his beef from the Midwest to local markets. (Profitable production of beef year-round from an operation as big as Tom's was not possible given the lack of facilities in Montana, but selling select cuts of his beef to local restaurants and processing less expensive cuts, such as roasts, elsewhere held promise.) Despite his laser focus on bovines, his deep care for the land and his family's strong tie to it guide his actions. His concern for the land he once owned is still evident as we walk through pastures and look into the cricks today. He remains proud of the grass this land can produce—each of three annual cuttings nets up to four tons of forage per acre. And his fascination for the full web of life that thrives in the waters, grasses, and big sky of the ranch is like that of a child seeing it all for the first time. He takes pleasure in knowing that his son can continue the family legacy, even if someone else now owns the ground. With John, he will continue to guide land use on the ranch and offer suggestions when asked by the out-of-state landlord. The Milesnicks are patient, and I suspect that their views will once again influence stream and cattle management along Smith, Thompson, and Ben Hart cricks. When the reed canarygrass forms a dense canopy over the sacrosanct waters, reducing their flow and stifling nutrient inputs that the trout rely on, the suggestion to graze them hard and fast will make more sense to the new landowner much as it has to others who once thought of cows not as tools but only as evil things.

PRODUCTION IS THE mainstay of a working ranch, such as the MZ Bar, but even here I saw nearly every acre also being shared with the deer, birds, and fish that abound on this land. Moreover, because grazing was done for only a few hours each month in the paddocks that encased miles of spring creeks, these areas were mostly spared for fish and wildlife. This combination of sharing and sparing produces a rich mix of habitats that support a wide diversity of native plants and animals while also delivering substantial revenue from the sale of cattle, hay, and grain. Using cattle to work the land benefits the local ecosystem by keeping invasive weeds in check and providing the economic return needed to maintain expansive grasslands. High-production hybrids, such as Garrison creeping foxtail, replace native bunch grasses on the ranch, but even in this prairie of our making birds such as bobolinks, savannah sparrows, western meadowlarks, and sandhill cranes thrive. The sheer size of the ranch afforded these animals a bit of refuge from the rapidly increasing human population just a few miles away.

Ranchers often perceive conflicts between their craft and the wildlife it may attract. In Montana, for example, bison are restricted from migrating beyond the boundaries of Yellowstone National Park because ranchers fear that they will infect their herds with brucellosis. This disease, obtained initially by bison *from* cattle, causes first-time pregnant heifers to abort their fetus. Bison were not an issue at the MZ Bar, being as it is about one hundred miles from the gates of Yellowstone. But deer were a bit of a curse to Tom. The several hundred white-tails that eat his grass and alfalfa were in his words "a thorn," but these too provide a valuable resource to local hunters who harvest fifty animals each autumn from the ranch. The benefits of allowing wildlife to graze among the livestock may extend beyond the simple provision of game. In central Kenya, for example, studies of livestock ranches that invite, rather than exclude, zebras and antelope have demonstrated increased weight gain by cattle that shared pastures with wildlife relative to those that grazed alone. This benefit may result from increased growth of plants in response to many grazers. The advantages of commingling grazers flow both ways. Ticks are diverse, abundant, and implicated in the spread of several deadly diseases in Kenya. Cattle are treated with insecticides to reduce the presence of ticks, and the efficacy of these treatments extends beyond the livestock. High-intensity grazing of pastures with treated animals resulted in significantly reduced overall tick abundance and the pathogens they spread, which benefits wildlife as well as humans.

Although maintaining grasslands, even those grazed by cattle and other stock, benefits a wide range of birds and other wildlife, the long-term consequences (and vulnerabilities) of grazing to climate change are becoming increasingly apparent. The removal of terrestrial plants by animals grazing natural grasslands and the conversion of forests to pastures for livestock together roughly halve the amount of carbon that can be stored by Earth's terrestrial plants. This undercutting of the global carbon cycle reduces the planet's ability to absorb and sequester the excess carbon dioxide we pump into the atmosphere. On northern ranges, such as those worked by the Andersons and Milesnicks, climate models predict warmer and wetter conditions in the future. This change may increase the productivity of grasslands to the benefit of the livestock industry. But the improved climate is also likely to foster increases in the spread of noxious weeds and pests that may challenge the health of cattle as well as humans. It is clear to me, in talking with ranchers in Montana and reading about their innovative approaches in Africa and elsewhere, that although grazing is undoubtedly part of the problem, using it as a tool in ecosystem management can also provide novel solutions.

THE MILESNICKS' LOSS of the MZ Bar is a tragic reminder of the vulnerability of farms and ranches to the whims of their owners. When farmers such as those we've met throughout this book own the land they love, work, and nourish, its fate is secure. To them, leaving the area or selling out is a refusal of their heritage. It is a last and painful resort. However, when those who have lost the personal connection to the property are responsible for deciding its fate, then greed and a desire for short-term gain increase the likelihood that the highest bidder will buy great lands with little concern for their future. Although the new owner of the MZ Bar has not fully embraced the land ethic that Tom built over his forty-some years walking the valley's cricks and uplands, he at least is conservation-minded. With John at the helm, the land will be well cared for, and if the current restriction of cattle from the streamside grasses proves unworkable, then their return may more definitively prove the worth of cows as tools.

IT IS HARD TO IMAGINE cattle ranchers more deeply connected to the land than the Milesnicks and Andersons, but I found one in Costa Rica who just might fit the bill. This renegade does not share his property with wild animals; he has turned it over to them. And in the process, he has given visitors from around the world a close look at the wildlife that convinced him to get out of the cow business entirely.

ELEVEN

A Cattleman Turns
Conservationist

A SOFT BREEZE FROM THE PACIFIC OFFERS little relief from
the midday sun that bears down on a soft-spoken, seventy-three-year-old
cattleman. I stand close to hear Jack Ewing tell me he dreams of tapirs. In his
subconscious, the beasts lumber along the trails he has created and munch a
salad of native plants he has helped to nurture. This vision is not the usual
dream for someone who began his career in cattle pastures scratched out
of some of the world's most productive forests. Tapirs shun pasture and require vast expanses of tropical rainforest, such as the Osa Peninsula's Corcovado National Park. They once roamed Jack's land, but since early cowboys removed nearly half of the ranch's eight-hundred-plus acres of forest
in the early 1970s, few have been seen. Even the hope of seeing a tapir would
have vanished if not for a chance encounter with a wild monkey that convinced Jack more trees would benefit the land. As these trees brought wildlife into more frequent contact with Jack and his family, their appreciation
for wild creatures and the forest they require grew. Jack's mind evolved,
and dreams of trekking tapirs soon replaced those of fat cattle.

Jack, his wife, and their four-year-old daughter flew to Costa Rica
on a DC-6 airplane with thirty-seven cows from Miami in 1970. Fresh out
of Colorado State University with a degree in beef cattle production, he
was eager to make a living ranching the rainforest to produce burgers and
steaks for hungry American mouths. He and his young family relished the
adventure of living on a farm in primitive conditions. Electricity did not
reach the ranch until the 1980s. Jack recounts these times as "the happiest
days of my life," musing how the family sought entertainment in the most
mundane of affairs. Cockroach control, for example, became a family event
as each Ewing hunted the pests with a plastic pistol that shot small discs.
Radio served as their only conduit to world affairs. Just getting to school
was a physical test because the bus with its shabby brakes could only slow,
not stop, as it passed the farm, so the kids (a son was born shortly after the
Ewings began ranching) had to run and jump to get aboard.

Opposite page: Baird's tapir once roamed Hacienda Barú and is hoped to do so
once again.

Raising cattle was big business at that time. Beef, coffee, and bananas were Costa Rica's top three exports, driving the economy and the clearing of land. Government policies enabled the expansion of agriculture, linking it to development, while international loans incentivized farming and especially cattle ranching. Cattle ranching benefitted more than grain farming or forestry in the 1960s and 1970s for three reasons: the demand for beef increased, loans could be secured using cattle as collateral, and ranchers could walk their cattle to market from roadless areas. As a result, pastures increased from 2 million acres in 1950 to 5.4 million acres in 1984. Overall, in Costa Rica forest cover dropped from 72 percent of the total land area in 1950 to 56 percent in 1970, and it reached a nadir of 26 percent in 1983. According to Jack, nobody thought about conservation back then. They just cut the forest down to make the land more productive. In short order, Jack was managing the cattle on two ranches, including Hacienda Barú, where Jack and the monkey were soon to meet.

As is customary in the tropics, Jack and his cowboys delineated their pastures with living fences: pickets of chest-high posts cut fresh and staked into the wet ground. Barbed wire strung between posts keep the cattle in, but the posts soon take root and grow into sizable trees, in some cases intertwining into living hedges. Crews planted stakes of thorny poro, peeling-bark gumbo-limbo (also called jiñocuabo), and madero negro single file at Hacienda Barú. In two quick years, the fence posts sprouted into trees twenty or more feet high. As the fences grew, they formed narrow corridors that outlined hundreds of acres of grass. The forest wildlife, including sloths, monkeys, pacas (large rodents), and opossums, often traversed the living fences.

In the early 1980s, Jack noticed a young male white-faced capuchin monkey making its way from the mangroves to the primary upland forest. To move between the two forests, the monkey navigated a rather sparse living fence. It stayed in the fence canopy when it could but often had to run awkwardly on the ground between the posts. In the trees, four hands and a prehensile tail allow monkeys to outmaneuver most predators, but on the ground they are vulnerable. One of the cowboy's dogs, Lobo, must have sensed this vulnerability, and it gave chase to the seven- or eight-pound pri-

A white-faced capuchin monkey.

mate. Just as the dog closed in for the kill, the capuchin leaped onto the dog's back, dug its hands into the mutt's eyes, and started biting the stock dog's ears. The dog had a heck of a time shaking the monkey, but it finally did, after which it tucked its tail and sprinted back to the stunned audience of cattlemen. Lobo never hunted again. Jack decided then and there to widen all the fences at Barú, making them out of three parallel rows rather than the usual single row. Doing so between the mangrove and primary forest would remove only two acres of pasture while allowing traveling monkeys to stay in the trees for both their protection and that of the ranch dogs.

After the monkey incident, Jack thought about other ways to simultaneously provide for wildlife and cattle. An easy first step was to quit clearing unproductive land that was too steep for cattle to navigate. These areas were hard to work and little used by cattle, so taking them out of production saved money by reducing labor costs. He started with seventy-five acres on a recently deforested hillside in 1979. In just a few years, the jungle came back with a vengeance, and wildlife became abundant and visible. Jack and his young family enjoyed seeing peccaries, monkeys, and toucans pass by their farmstead. They mostly just let nature recover, though they installed a few small teak plantations. Their interest in conservation forestry took off when they learned about the significant decline in ceiba trees (also known as kapok), a majestic species with a trunk that rises hundreds of feet before branching to emerge from the high rainforest canopy. To help the ceiba, they planted some seedlings in the recovering forest.

Whenever I visit Barú, I check the ceibas that Jack planted. One is well on its way to dominating the canopy right next to the restaurant. It soars ninety feet and has a canopy that blots out a forty-foot-wide hole in the sky. Its roots are simple flutes from the trunk, which is arrow straight and tubular for eighty feet before bifurcating and branching into the characteristic umbrella-like top. I sidle up to the base to measure the diameter of the trunk, which exceeded three feet in 2017. The cottony seeds of the tree can be gathered to make natural insulation, which a few decades ago was often used to stuff mattresses, couches, and sleeping bags. Today it is an ecofriendly filling favored for pillows and meditation mats. It can be purchased online for nearly ten dollars a pound.

Jack's ideas continued to evolve as he observed the forest and read about its importance to the ecosystem. A cowboy who brought him an ocelot that he had shot on the ranch threw Jack into a moral dilemma. He thought that the skin would be a nice addition to his wall, but he couldn't live with the thought of aiding and abetting the killing of the last forest cat of Barú. He reconsidered the local custom of allowing hunting anywhere at any time and banned hunting at Hacienda Barú. He hired guards to protect the forest wildlife (most hunters were after the delicate meat of the paca, which frequents Barú and is the prime prey of forest cats). The locals thought he was crazy, especially when he started catching rogue hunting dogs on the property. He'd take the dogs to the jail in the nearby town of Dominical, where the hunters would have to sign a confession of trespassing and illegal hunting if they wanted to retrieve their hounds. Then they had to promise never to hunt on Barú again. Jack kept the disclosure rather than pressing charges, but he made it clear that if the hunter broke his promise, the confession would be handed over to the police for action. A prosecution was never needed, as no hunter dared go back to Barú. Despite his evolving land ethic, Jack did not consider himself a conservationist. "I was a cattleman," he tells me, but "the locals thought of me as a conservationist."

By 1985, Jack finally accepted that he was a conservationist. He enjoyed showing neighbors and visitors the monkeys of Barú. Visitation increased in this part of Costa Rica because of improved roads, bridges, telephones, and electricity. Across the country, this was also a time when forest owners saw their lands as valuable assets in the emerging business of ecotourism. At first, Jack did not charge for his tours, but some visitors insisted on paying. He made three hundred dollars his first year, fifteen hundred dollars the second, and fifteen hundred dollars in a single month of the third year. In 1990, as the beef export business began to collapse, Jack and the partners with whom he owned Hacienda Barú sold off all the cattle. His six partners looked to sell their share of the ranch. An American businessman with interests in international environmental education bought out the partners, and he and Jack quickly set about fully converting the cattle ranch to an ecotourism retreat and wildlife refuge. They built guest cabins, a central dining pavilion, and a swimming pool. Harvesting teak from their plantations supplied all the lumber required. They hired and trained an expanding staff of nature guides, cooks, maids, guards, and office workers. And they let the forest reclaim an ever-increasing portion of the ranch.

By 1997, the once fragmented rainforest was again continuous, from the highlands to the coast. The expanded forest cover is evident on the aerial images from 1972, 1997, and 2005 that Jack proudly shows all visitors. Looking at them you can see especially the reforestation of pastures on either side of the Barú River. Jack now manages the natural habitats — rainforest, mangroves, beach, and river of Barú — as a nationally recognized "Private Wildlife Reserve." Technically, the reserve is a mixture of private and public land. In Costa Rica, all the area within a bit more than two hundred yards of the average high tide line is public property, and this includes some of Barú's beaches and mangrove forests. Jack's partner holds title to

An aerial photo of Hacienda Barú in 1972. The ranch encompasses the area to the left of the Barú River (the river mouth is visible just to the right of the bottom, middle of frame) along the coast and into the adjacent mountains (forested area above road that contours from river to left). The land along the coast and into the highlands has been cleared for cattle grazing at this time. (Photo courtesy of Jack Ewing)

Hacienda Barú in 1997. The plume of sediment from the river is evident into the Pacific. Living fence lines now border the pastures and connect the upland forest to the newly established coastal forest. (Photo courtesy of Jack Ewing)

808 acres of the reserve. Jack maintains ownership of the tour company, exclusive rights to take tours into the forest (for which he pays 25 percent of the gross tour income to his partner), and seven acres that contain six cabins, seven hotel rooms, a zip line, gardens, a gift shop, and supporting facilities for tourists and staff.

The investment in conservation and ecotourism has proved economically and socially sustainable. Jack's tour company broke even during the economic downturn from 2008 to 2015 that saw international travel decline worldwide. In 2016, Hacienda Barú hosted seventeen thousand visitors and made a significant profit. Visitors can now choose from among four guided

Hacienda Barú in 2005. Forest now dominates the ranch, creating an important refuge and connection to more extensive forest in distant mountains. (Photo courtesy of Jack Ewing)

bird-watching tours, four guided hiking tours that emphasize cultural and natural history, a guided overnight camping tour to the upper rainforest, a guided zip line tour, and a guided climb into the canopy, or they can explore well-maintained trails on their own. Jack's employees are well cared for and remain at Barú for many years. The tour company employs thirty-nine local people, ten times what the ranch once supported. Each, whether maid, guide, or office manager, owns their own transportation—a motorbike or car—that gives them freedom and mobility away from the job. His guides are encouraged to train with national experts and become certified in transferable skills such as emergency aid. Carlos Jiménez has guided guests around Barú for a decade. He is proud of his country's wildlife and loves sharing his passion for the forest with visitors. As I walk with him, he lists the benefits of working for Jack: a good wage, overtime pay, health care, child support, and a guarantee of year-round employment. His only lament is the long hours, which in part cost him his marriage. His schedule provides less time for parenting his young son than he would like, but he is grateful to be able to afford the monetary support his child needs.

Employees such as Carlos know that their job depends on the rainforest. And it seems that the local community also has learned this lesson. Jack often tells a story about a hunter who chased him with a machete. Jack had captured his dogs after the ranch was closed, and the hunter did not want to sign a confession. Recently the hunter's son, a local taxi driver, dropped guests off at Barú and made a point of telling Jack that he now understood why the rainforest was important. His job, and those of his seven siblings, all depended on ecotourism. He was appreciative. His father, the old hunter, however, refuses to make the connection. He hasn't pulled out the machete lately, but he won't acknowledge that a tourist paying to see wildlife has provided more sustained benefits to the local economy than have hunters who would harvest the wildlife.

The value of Barú's 815 acres has steadily increased despite—or perhaps because of—its evolution from a cattle ranch to a tourist destination. Jack and his original partners purchased the property for $100,000 in 1972. The partners sold their 87 percent stake in the company in 1993 for $1.4 million. In 2017, the value of Jack's 13 percent stake alone was $1.4 million. An interested buyer tendered an offer of $15 million to buy the entire 815 acres with no strings attached in 2016. Much of this value is due to the business opportunity provided by the wonderful forest Jack created and preserved. Jack and his partner declined the offer, fearing that the refuge they built would succumb to further development.

AS A MIXED-SPECIES flock of birds envelops me, I appreciate the natural capital that Jack built into Hacienda Barú. I've climbed atop a small tower available to visitors just beyond the lodge. Scarlet-rumped tanagers squeak and dash about in their velvet black and crimson plumage. A barred woodcreeper probes the moss and bromeliads for insects. A line of leaf-cutter ants passes by, taking leaves and flowers to their underground fungus gardens. I spot a robin-sized unfamiliar bird moving slowly, sallying forth occasionally as it works the middle branches of the large strangler fig trees. Its drab rufous and greenish plumage and actions remind me of a flycatcher, but I cannot get a clean look. A second bird, this one with a distinctive black cap and white streaks on its shoulder and flight feathers, is also unfamiliar.

I snap a quick picture so that I might later identify them. (They turn out to be my first look at white-winged becards.) The parade of birds continues for several minutes as I encounter a lineated woodpecker—a species that always reminds me of the pileated woodpeckers that roam my home forests near Seattle. Trailing the main flock is a familiar migrant that might have traveled from a forest near my Washington home, a warbling vireo. I end the excitement watching a moderate-sized hummingbird with the fantastic label of violet-crowned woodnymph. The birds move above, below, and beside me as I wonder aloud if this really could have been cattle pasture a few decades ago. It seems impossible.

The ecological sustainability of Barú is undeniable. The 815-acre forest is now part of a regional corridor designed to connect the various mountain and coastal forests among three national parks (Corcovado, Manuel Antonio, and Los Quetzales) so that the largest forest denizens might again walk the hills and beaches. The area, known as the Path of the Tapir Biological Corridor, includes two hundred thousand acres—nearly 250 times larger than the area of Hacienda Barú. Because of restoration and continued stewardship, over half of the corridor is a primary forest (never-before-cut rainforest) and secondary forest that has been untouched for at least forty years. Within the corridor are twelve thousand acres of commercial farmland and fifty-two communities. Though tapirs and jaguars are not yet resident in the corridor, some 1,000 species of plants, 146 species of mammals, more than 600 species of birds, 81 species of reptiles, and 51 species of amphibians do live here.

My students and I have been fortunate enough to get intimate looks at some of the larger mammals that use the trails at Barú. Coatis, capuchins, and agouti regularly work the old chocolate orchard. Sloths, both three- and four-toed species, hang out in the cecropia trees that are common in the young forests leading to the more mature uplands. Dog-sized anteaters, known locally as tamanduas, use their strong claws to rip into the abundant arboreal termite nests that decorate the forest like large brown ornaments. Neotropical river otters sometimes poke their heads above the thick grass that covers a wet pasture along the dirt road to the beach. Collared peccaries are always exciting to see. Mostly they grunt, release a musty scent,

A shy riverside wren sings from the understory.

and scurry off. They are known to charge a person, so we are always alert when we surprise them. I was alone on the trail during my last afternoon at Barú, engrossed in photographing a well-lit riverside wren as it sang forth, when I felt the presence of something behind me. As I turned to look down the trail, I saw two peccaries sauntering right toward me. They munched wild plums that littered the forest floor, ambling on and off the path, until they were only a few yards from me. As I took their pictures, the large male and smaller individual casually wandered off. They never were startled and seemed comfortable in my presence, if they even noticed me. I replay that memory often, as I suspect other visitors also do of their favorite moments in this section of the Path of the Tapir.

Jack hikes up the mountain trail to his zip line every other day. The forty minutes of exertion keep his heart healthy since the installation of a pacemaker three years ago. It also allows him to keep tabs on some of the 369 species of birds and 85 species of mammals (24 of them bats) he's documented on the ranch. I suspect that his eyes scan every muddy patch for the distinctive three-toed track of the tapir. He looks for a sign of the tapir more widely as well, deploying eight camera traps (motion-sensitive cameras that photograph animals passing by day and night) throughout the hacienda. Jack's cameras have recorded many of the common, and several noteworthy, animals, including puma, ocelot, and jaguarundi. They are still waiting to get the first photo of a tapir.

A shared vision of ecological, social, and economic sustainability guides land management throughout the Path of the Tapir. Indeed, Costa Rica is a world leader in forest conservation and ecotourism. In 2011, Costa Rican president Laura Chinchilla noted that her country was proud to have reversed its deforestation rate, which had dropped from a peak in the mid-1980s at 124,000 acres per year to no net loss in 1998. The plants have responded as roughly half of the country is now forested. Thanks to forest restoration, Costa Rica is a hotspot of biodiversity, worldwide, but unlike many similar places, it is not also a cauldron for extinction. The International Union for Conservation of Nature keeps track of the extinction and endangerment of plants and animals on its Red List. One of the riskiest nations on that list is the United States, where our actions threaten 14 percent of 8,812 assessed plants and animals with extinction and 283 species (3 percent) are known to have gone extinct. In Costa Rica, human action threatens 8.7 percent of assessed plants and animals, and only 3 species are known to have gone extinct (though assessments exist for less than 5 percent of the 9,000 species of plants).

President Chinchilla suggests that an amalgamation of several national policies have accomplished the reversal in forest cover: a bold decision to disband the country's army in 1948, decisions to use payments for environmental services (funds from the nation and international community to reserve forest) to reduce rural poverty, and the creation of a federal Forestry Department. These decisions have paid off for the country as a whole, which now gains half of its gross domestic product from tourism, especially ecotourism. However, threats to the forest continue to evolve, today coming mostly from expanded development. Around Barú, the population in the small surf towns of the region is growing, and aging residents such as Jack are retiring. The fate of Barú's forest and Jack's dream of tapirs walking its trails may well hinge on the land ethic of the

country's leaders as well as that of the next owner here and throughout the
Path of the Tapir.

REFLECTING ON HIS life's trajectory, Jack emphasizes that none of this was
planned. He never knew where life would take him. In the early 1970s, Jack
recalled, "setting up a nature reserve was the farthest thing from my mind."
He did not know what biodiversity was. (That term, a contraction of *biological diversity,* wasn't coined until 1985.) How does a specialist in beef
cattle production pinball through life and wind up a conservationist?

I see Jack's evolution as a rancher as akin to other forms of organic
evolution. Change, in a flower's corolla, bird's tradition, or rancher's mind,
can take unanticipated paths because the forces of nature do not work
toward some apparent goal. Current morphology and behavior evolve to
fit the prevailing environmental conditions that challenge a parent. Jack
did exactly that when he responded to a monkey's plight by planting more
trees. Then, as he began to benefit from the increase in forest, he adapted his
skills as a land steward in a new direction to guide people rather than cows.
His ability to change benefitted his family. The culture he created is now
filtering across generations. Although his wife still misses the cows, their
son and daughter have just returned to Barú, where they will comanage the
ecotourism company that Jack built. This cohort of Ewings brings skills in
tourism, architecture, and business administration to the ranch. I expect
that these stewards will listen to the land—the wildlife that lives on it and
the staff that tends and interprets it—to strengthen the positive feedbacks
among its ecological, social, and economic conditions that Jack re-created
when he evolved from cattleman to conservationist.

JACK'S STORY UNDERSCORES how quickly land and those who work it can
change their character. In just over a half century a forest transitioned to
pasture and back to forest. The capacity of agricultural land to change—to
be restored to a more natural state or to be converted to a new housing development—is essential to remember. As the world population continues
to grow, few individuals may have the luxury to rewild a farm: economics
and societal norms may preclude it. But, as Jack experienced early in his

transformation, even small changes to an intensively ranched area, such as widening fencerows or allowing nature to reclaim unproductive land, can provide benefits for wildlife and the rancher. A willingness to let nature onto the ranch is catching on in parts of the American West, as we saw on Tom and Mary Kay Milesnick's MZ Bar Ranch and Hilary and Andrew Anderson's J Bar L Ranch. With incentives, even some Nebraska corn farmers leave a little grass for meadowlarks. But in some settings, a change in the character of the cattle ranch and cattle rancher remains unlikely.

Anthropologist Jeffrey Hoelle spent eighteen months with ranchers in the Amazon of Brazil to understand their personas. In the early 1970s, the Brazilian government encouraged the urban poor, small landowners, and landless people to homestead Amazonia. Large-scale cattle ranchers and speculators were also attracted to the region at this time—just as they were to Costa Rica—because of global demand for beef and government incentives and subsidies that enabled them to pay little for large rubber plantations. The cattle ranchers of Amazonia, whom many characterize as elitist white men from outside the region, are infamous for inciting violence, displacing native people, and destroying the rainforest. Hoelle discovered that the truth was a bit more complicated. Some ranchers did come to the region already possessing great wealth and influence. They bought large swaths of land and accrued herds of up to fifty thousand head. Others went to the rainforest from much more humble origins and gradually increased their landholdings through hard work. Despite these different beginnings, all the ranchers share a strongly pragmatic view of the land. As in the Bible, to them, the function of land is to provide food.

Social status and a man's character in rural Amazonia are affected by one's ability to transform the land so that it might provide food. Manliness equates to the transformation of the land. "Individuals who do not show clear separation of human and natural spaces are considered to be lesser men regarding gender roles, which assert that real men should perform physical labor." Here "a property with well-maintained, clean pasture demonstrates that the owner is a hard worker." In contrast, a person who barely distinguishes his house from the forest is considered lazy or unwilling to work. Some may even be considered subhuman. During his interviews,

Hoelle found that "work and landscape transformation also serve to distinguish, in the minds of some, men from less evolved people and animals." With such strong beliefs, it is no wonder that Amazonian cattle ranchers feel unjustly regulated by new governmental policies that aim to conserve the forest. Such thinking constrains their manhood.

Rural culture, in which farming and ranching are embedded, greatly influences the evolution of agriculture and its effects on wild creatures. The view that land was a commodity meant to feed humans was fundamental to homesteaders and early agrarians throughout the world. Nature was in the way, and people needed to conquer it. To various degrees, I suspect that all farmers and ranchers hold on to this utilitarian, anthropocentric view of the land. The opinions of Amazonian ranchers, for example, are extreme in this degree. I also felt it strongly among the grain farmers of Nebraska. But I feel it to a much lesser extent among the diversified small-scale organic farmers I've met in Washington and Costa Rica. And in Montana, progressive ranchers see the land as providing forage for cattle and habitat for fish and wildlife. As society comes to value land for its multiple uses, these new farmers are evolving their land ethic to one espoused a half century ago by Aldo Leopold. Property to the new agrarian is both a commodity and a community. But although even progressive farmers and ranchers recognize the community on their lands, their livelihood depends on its ability to produce a useful and sustainable commodity. Jack Ewing's ethic has emphasized the land as a community, but in so doing he has discovered another way in which it provides a durable commodity. No longer does his land provide food. Instead, it contributes to global conservation efforts and feeds our desire to know the others with whom we share the planet.

REDUCING AGRICULTURE IN some parts of the world may seem at odds with the goal of feeding a growing human population. However, if we are to nurture people and nature in the future, farms — and forests — will be needed. Trading cattle pastures for forest seems reasonable, especially in the species-rich tropics. As discussed in chapter 9, reducing the amount of meat we consume will increase the ability of future farmers to feed a world of nine to eleven billion people. A 2017 study emphasizes the importance of focusing conservation efforts on the world's largest intact forests. After assessing the status of nearly twenty thousand vertebrates on the Red List, Professor Matt Betts and his colleagues estimate that failure to curb the conversion of now-extensive forests in hotspots of biological diversity, such as the western Amazon, Borneo, and the Congo Basin, will increase the odds that species there will decline in population size and be at higher risk of extinction. They conclude that "new large-scale conservation efforts to protect intact forests are necessary to slow deforestation rates and to avert a new wave of global extinctions."

Slowing deforestation in the Amazon hinges on changing the way owners ranch their cattle. Regulations vary by country, but many intend to reduce forest loss on ranchlands. In Brazil, for example, ranchers are required to maintain 80 percent of their land in forest. However, despite environmental regulations, cattle ranching remains the most important driver of deforestation throughout the Amazon. In Brazil, the Red List classifies 15 percent of species as threatened, and eighteen animals and plants are already known to have gone extinct. Increased enforcement of regulations may stem forest loss on cattle ranches, and certification of beef as coming from forest-friendly sources can incentivize good stewardship. But for many Amazonian cattle ranches to truly contribute to conservation, a significant change in ranchers' attitudes will be needed. Greater understanding of the economic value of ecotourism might start to broaden their land ethic. Only then will a forested pasture increase the social status of its owner. As long as a clean pasture equates to a rancher's manliness, society will marginalize forests.

MY STUDENTS ARE packing the bus as we prepare to depart Hacienda Barú. Just before we leave, a commotion in the large mango tree behind Jack's office draws our attention. A troop of capuchins is on the move. They drop singly from the tree and adjacent wire to sprint across the lawn behind the cabins. They are alert and vigilant as they travel on the flat ground. Babies ride, like jockeys, on their mothers' backs. In a moment they are back in the safety of the trees, which now stretch across old pastures to the sea. They relax, search for fruits, lizards, and insects. Jack smiles and wishes us well.

His face is aglow knowing that his transition of the land has provided fifteen young adults from Seattle and a troop of monkeys from Barú an opportunity to understand one another better. We—and the other seventeen thousand visitors who came here in 2016—have come to know and respect Jack as a hard worker with pastures that are no longer clean but instead dirty with life.

JACK IS NOT ALONE in evolving in response to his agrarian life. Many plants, insects, microbes, and animals have done the same.

JUST OVER A CENTURY AGO, FLOCKS OF Carolina parakeets (*Conuropsis carolinensis*) decorated the forests of the southeastern United States with their brilliant green, orange, and yellow plumage. Their beauty must have brightened the drudgeries of American rural life, but their chittering flight calls were a death knell to farmers. The parakeets flocked into cornfields and ate the precious grains. In response, farmers shot the crop raiders in large numbers. Killing, along with clearing of forests where the birds nested, spelled doom for the once superabundant species. A few may have existed in South Carolina through the late 1930s, but the last official record came from the Cincinnati Zoo in 1918. In that year, a pair of the yellow-headed beauties, named Lady Jane and Incas, finished their long lives—more than thirty years—just a few cages away from where the last passenger pigeon died four years earlier. Incas was the last to go, dying a few months after his lifelong mate. His keepers, noting the pair's strong bond, cited grief rather than old age as the cause of death.

Agriculture is thought to have caused the extinction of nearly thirty species of birds, including the Carolina parakeet. The International Union for Conservation of Nature documents 156 modern bird extinctions on its Red List. That is just over 1 percent of all birds known to have existed in the past two thousand years. Human action causes virtually all bird extinctions—most from our introductions of nonnative predators and diseases to fragile island ecosystems. Overharvest is also a prime driver of extinction. But on continents, the main reason birds go extinct is from the clearing of land for agriculture. Farm clearing, in part, doomed the Carolina parakeet. Our transition to agriculture also pushed passenger pigeons (*Ectopistes migratorius*) in North America, paradise parrots (*Psephotellus pulcherrimus*) in eastern Australia, and laughing owls (*Sceloglaux albifacies*) in New Zealand toward eventual extinction.

When our agricultural actions do not fully extinguish species, they

Previous page: Great tinamou, an early evolved species, are difficult to maintain outside of large forest reserves.

are often locally extirpated. In my casual strolls among monocultures of corn, soybeans, wheat, pineapple, coffee, or palm oil this local impact was all too apparent. In these simple and rapidly changing landscapes extirpation claims many species that depend on stable, native ground covers, including tinamous, curassows, pheasants, and meadowlarks. Gone too are those with specialized diets. Because many of the missing species on agricultural lands are from lineages with few local relatives that often evolved early in Earth's history, their removal dramatically reduces the evolutionary legacy maintained on farms. This simplification of bird communities, regarding the number of species as well as their evolutionary distinctiveness, does not bode well for the resilience of these assemblages or their ability to provide people with services, such as pollination, seed dispersal, pest control, and cultural stimulation.

Scientists from Stanford University and the University of California surveyed birds in Costa Rican forests, crop monocultures, and wildlife-friendly farms for twelve years to reveal the impact of extinction. They found an average of sixty-two types of birds in forests, which represented 4.1 billion years of evolutionary history (the sum of the estimated time since each species diverged from its closest relative). Monocultures only harbored half as many species as found in forests, and because the majority of extirpated species were distinct and ancient, these intensive agricultural lands lost 900 million years of the avian evolutionary history contained in forests. Wildlife-friendly, diversified farms such as those I visited in Mastatal and Piro were nearly as diverse as nearby forests, but because recently evolved songbirds, such as sparrows, seedeaters, and blackbirds, replaced more ancient and unique trogons and tinamous on the farms, they lost 300 million years of evolutionary history relative to forests. Scientists have also documented such losses on French and Italian farms. In France, recently evolved skylarks and corn buntings thrive on intensively farmed lands, but the loss of overall evolutionary history is in direct proportion to the degree of agricultural intensification. In central Italy, where the European Union incentivizes the maintenance of low-intensity, traditional agriculture, such farms maintained 20 percent more bird species and 27 percent more avian evolutionary history than did high-intensity monocultures. Around the world, those species that evolved most recently, especially those that exploit seeds or flying insects, readily adapt to farm life, while the agrarian setting quickly excludes ancient ground dwellers, fruit eaters, and predators.

While we grieve the loss of species, we should also celebrate those that can thrive in our shadow. For me, seeing a family of horned larks in a fallow field of wheat stubble or discovering that an unfamiliar chatter within a plantation of oil palm comes from a variable seedeater is a reminder to applaud the successful, often recently evolved, species among us. This counterpoint to extinction—the evolution of adaptations that quickly buffer some birds from our belligerence—is a poignant demonstration of the power of natural selection to fit some organisms rapidly to hostile environments. Most adaptations involve changes in behavior, but some also come with fundamental shifts in an animal's physique. These evolutionary changes occur over decades or centuries, not the vaster geologic time scales we often think of as necessary for natural selection to fiddle with an organism's characteristics.

HUMANS ARE ONE OF the most significant evolutionary forces affecting the organisms that share the planet with us. Our incidental or purposeful feeding of birds has enabled new travel routes to evolve, for example in some blackcaps (a type of warbler) that now migrate from German breeding areas to British bird feeders instead of North African olive groves for the winter. Our noisy traffic favors birds that sing louder and at a higher pitch so they can get their messages through the anthropogenic din. Fishing for our food drives down the size of fish so that they can slip through our nets. The medicines we use to stave off infection select for new, more virulent, resistant diseases. In the same way, our agrarian lives offer wild species new opportunities and challenges to which they evolve.

As its name suggests, the short-toed snake-eagle subsists mainly on a diet of reptiles. In the Dadia Forest of northeastern Greece, for example, these raptors regularly eat eight species of lizards and ten species of snakes. But in recent years their diet has changed to include the bonanza of voles, jirds (gerbils and their kin), and mice displaced by deep plowing of wheat fields that surround their breeding areas. Deeper tilling exposes more rodents and attracts more snake-eagles to the grounds. The shifting of snake-eagle diets to capitalize on the bounty of agriculture gives Greek farmers a bit of pest control and sets the eagles on a new evolutionary course that may profoundly change their character.

In northern Cameroon, another raptor has shifted its diet in response to agriculture. The grasshopper buzzard, as you might guess, typically eats grasshoppers as well as rodents in natural savanna habitats. But in agricultural settings, this hawk feasts mostly on abundant lizards. This shift might not be so good for the birds. Their nestlings grow more slowly and obtain subpar weights on the farm-to-nest diet of reptiles. So although a shift in the food may allow some buzzards to subsist on farms, they will need further refinements to excel there.

In contrast to diet, the migratory habits and routes of birds may seem immutable. The timing, distance, and direction of migration in many birds are under genetic or cultural control and therefore readily adjusted by natural selection. Many species, such as the blackcaps mentioned earlier, reduce the distance they migrate as foods become abundant in cities or agricultural fields along their path.

Whooping cranes provide a farming example. These endangered white birds with black flight feathers were reintroduced to the eastern United States in 2001 and learned to migrate between breeding grounds in Wisconsin and wintering grounds in Florida by following an ultralight aircraft. However, within fifteen years the majority of young cranes had shortened their annual migration to overwinter in the cornfields of Illinois and Indiana. This new tradition cut the distance traversed in half (from 1,100 to 560 miles). The spread of corn throughout central North America has similarly influenced the migratory routes of sandhill cranes, which stop to feed on waste grain in Nebraska as they move south from Alaska and Canada to overwinter along the U.S.-Mexico border. American crows likely followed corn crops north and west into prairie states and provinces and developed annual southward migrations as they consumed waste grains and winter froze the land. Movements are not the only aspect of birdlife to change with agriculture.

Farming the Imperial Valley of southeastern California has darkened the land as it irrigated and tilled historic light-colored desert. In response to life among darker soils, the region's horned larks have also become much darker in plumage. Larks collected from 1918 to 1934 and preserved in museums had feathers that were light sandy brown with little dark flecking. After eighty years of intensive agriculture, however, new specimens collected from 1984 to 2014 had rich brown backs and napes, increased dark flecking, and masks that were jet black. Dark birds likely originally came from those overwintering in the valley from darker subpopulations, such as the Channel Islands, because historical specimens did not include any dark or even darkly flecked larks. Regardless of the source, natural selection worked quickly to fashion larks better able to hide among the agricultural soils from keen-eyed predators such as northern harriers and prairie falcons.

Few birds have evolved so intimately with our agricultural ways as has the house sparrow (*Passer domesticus*). This familiar bird now inhabits portions of Europe, Asia, the Americas, South Africa, Australia, and New Zealand because of its ability to exploit our cities and farms. It feeds off our crops and waste and nests in our buildings and other structures. Genetic analyses conducted in 2012 point to a single origin of this commensal relationship, and as we might expect for this agrarian bird, the root is in the Fertile Crescent. There, ten thousand years ago, sparrows that had existed in natural grasslands feeding on seeds for hundreds of thousands of years became sedentary and fed on the grains our ancestors cultivated. As agriculture flourished in the Middle East and spread throughout the Palearctic and Eurasia, so too did house sparrows. As the sparrows encountered new climates and foods, their plumage and morphology changed—for example, darker and larger sparrow subspecies evolved in coastal and colder climes, respectively. When house sparrows followed humans across the Alps into present-day Italy some eight thousand years ago, they hybridized with local and closely related Spanish sparrows (*Passer hispaniolensis*) to seed the evolution of what would eventually become a distinct species known today as the Italian sparrow (*Passer italiae*).

Today one subspecies of the house sparrow (*Passer domesticus bactrianus*) retains its old habits and shuns human offerings. It breeds in the grasslands and wetlands of Central Asia, feeds on native seeds, and migrates to western India and eastern Iran for the winter. This wild sparrow remains isolated from surrounding sparrow populations that are commensal with humans by not interbreeding. And it is not evolving in response to recent changes in crops as are sparrows associated with people. During the modern agricultural revolution, the size and toughness of oat and wheat grains have increased throughout the range of commensal house sparrows. In response, sparrow beaks have become larger and more pointed. Because commensal sparrows are not migratory, selection did not favor streamlined, weight-saving physiques, and instead favored wide and heavy skulls that supported enlarged jaw muscles that increased the crushing capacity of the beak. Migratory house sparrows that do not inhabit human agriculture differ from commensals in having lighter beaks that are adequate for handling small native grass seeds and efficient to carry on their annual migrations.

Some of the birds that withstand our agrarian lifestyles come to intertwine their cultures with our own. As I write, my neighbors are readying for Halloween. They place pumpkins on their porches, drape fake spider webs over their shrubbery, and affix a scarecrow replete with model crows to their door. The evolution of the scarecrow into this—a holiday decoration—is the latest stop in a long coevolution between humans and the birds we have tried to keep from our crops.

Farmers and the crow clan have been in a sort of cultural arm's race for millennia. When crows started stealing our goods, be they salmon drying on the racks of Makah Indians in the Pacific Northwest, or corn growing in fields tended by members of the Mimbres people in New Mexico, we fought back. The Makah built scarecrows into their drying racks to ward off thieving ravens, while the Mimbres hung crows from fences surrounding their crops. Both strategies took advantage of the birds' abilities to associate danger with human food and therefore avoid it. A Mimbres bowl from the tenth or eleventh century captures this early point in our coevolution. Depicted on the pottery is a man facing a field of corn and holding three nooses high overhead. Similar nooses are spaced regularly along the fence around the plot of grain, each heavy with a crow hung by its neck. Behind the trapper, two live crows walk away and inspect another noose, as if learning from the experience. The use of dead crows remains an effective deterrent today. In a series of field and lab experiments, Dr. Kaeli Swift and I demonstrated that the sight of a dead crow motivated learning about the people and place associated with the victim as well as stimulated the parts of the crow's brain associated with learning, memory, and critical reasoning. As useful as a dead crow is at scaring its kin from our fields, it quickly became impractical. One alternative was widespread as early as the sixteenth century: the employment of young men as *crowboys* or *crow-herds* to chase crows from European wheat fields. Scampering after crows was also likely quite effective, but as with the use of dead crows, it too became impractical as our farming enterprises grew. The placement of stationary scarecrows replaced the crowboys, and then the real cultural coevolution began.

As crows habituated to the motionless scarecrows, farmers added moving parts or just moved them irregularly throughout their fields. Again, crows habituated and forced farmers to innovate further. Scarecrows today often include air cannons that fire at unpredictable intervals. Others fly high above the field as crazy-eyed kites. In some places, high-powered lasers are used to flush the birds. Each time as we develop a new scare tactic, the birds adjust and force our hand. At some point, our frustration at keeping crows and other grain-eating birds from our fields crept into popular culture. Today, a scarecrow is most reliably found on decorations symbolizing harvest time rather than in a field doing its original work. Our battles with crows stimulated our culture of pot-making, crop defense, harvest celebration, and storytelling. As Tony Angell and I concluded in our book *In the Company of Crows and Ravens,* scarecrows have "evolved out of the agricultural fields and into our art, literature, film, and fable."

Crows' culture has changed in at least two fundamental ways in response to our agrarian lifestyle. First, they incorporated grains, such as corn in North America, into their diets. This dietary shift brought them into close contact with farmers, who often persecuted the birds with the gun as well as the scarecrow. Second, crows adjusted their defensive behaviors to the harassment meted out by farmers. Because of safety concerns, crows rarely are shot or persecuted in urban areas, so the effects of harassment can be seen when we compare urban and rural birds. In doing so, Dr. Barbara Clucas and I found that rural crows flushed at a greater distance from an approaching human than did urban crows. Rural nesting crows and ravens also appear more suspicious of people than do nearby urban birds. If a person approaches a rural nest, the birds sneak off and avoid conflict. But in a metropolitan area, the disturbed parents often dive wildly at intruders, occasionally rapping them on the head. Urban and rural crows are following distinct cultural paths because of their interactions with people.

The changes in behavior, culture, and appearance of farmland birds over time scales we humans can appreciate offer tangible examples of the evolutionary process. But to grasp the rapidity and ubiquity of evolution in the lands we have so profoundly altered, we must take a closer look at how plants, insects, and microscopic organisms, those creatures with short generation times and easily mutable genomes, evolve on the farm. Understanding their evolution is fascinating—and sometimes downright scary.

FARMING AFFECTS THE evolution of wild species in three distinct ways. First, tamed species influence the evolution of their wild ancestors. Second, the actions of agriculture affect the evolution of pests we aim to control by confronting them with substantial challenges to their survival and reproduction. Last, by introducing new, domestic genomes into the environment and often transporting these across natural boundaries to gene movement, we affect the evolution of other wild species.

Wild plant evolution has been affected by all three of these factors. As a result, the speciation rate of plants is now higher than at any time since their initial development 700 million years ago. This burst in the creation of new species, which is now fifty to three hundred times greater than it was before the invention of human agriculture, is similar to bursts that followed other mass extinctions, such as when asteroids slammed into Earth 65 million years ago or when continental glaciers retreated at the end of the Pleistocene about 12,000 years ago. In Great Britain, for example, six or seven new, sexually reproducing species in four plant families originated from 1700 to 2010. This flurry of evolution occurred as wild species hybridized with domestics.

Most hybrids cannot reproduce sexually because the length, gene position, and shape of the parents' chromosomes do not match. Matching, or homologous, chromosomes, one from each parent, pair up and exchange bits of their DNA before being segregated into viable sex cells during the cellular process known as meiosis. However, nonmatching chromosomes, such as those held within the cells of hybrids between distinct species, do not pair up properly before exchange and separation, which fouls the ability of hybrids to form viable sperm and eggs. Plants, however, overcome this limitation by frequently doubling their entire set of chromosomes. In doing so, a hybrid plant not only gains a homolog for each of its distinct chromosomes but also attains an enlarged genome that is distinct from either parent. These differences in numbers and types of chromosomes enable the hybrid to sexually reproduce and remain unique even if in the company of its parents. In this way, several new daisy species evolved in Britain, and the cultivated radish hybridized with the wild jointed charlock to form the now-abundant California wild radish. When hybrids or other native plants confront the challenges of living within agricultural lands among domestic species, further evolution can occur.

Barnyardgrass is a hybrid that has evolved to look like rice. Its uncanny mimicry confuses farmers, enabling it to evade weeding. And this weed, one of the costliest in all of agriculture regarding reducing our harvest, also times its reproduction to coincide with that of the local rice crop. Some varieties even track subtle differences in farming practices, for example, being able to germinate in flooded or dry fields depending on the prevalent custom. Sometimes renegade individuals of one crop can invade another crop and evolve weedy attributes similar to those of barnyardgrass. Weedy rye, for example, which plagues wheat crops, likely originated from cultivated rye in California. Because weedy rye flowers later than cultivated rye, it has become reproductively isolated from its domestic progeny and has quickly evolved several unique traits that allow it to thrive in wheat fields. These adaptations include small seeds that easily shatter from the plant, much like those found in the native weed before domestication into cultivated rye. Remarkably, this evolution has occurred in just sixty years.

It seems that nearly everything the farmer does to reduce pests ac-

California wild radish blooms prolifically.

celerates their evolution. With the advent of herbicides in the late 1940s, farmers spurred the development of chemical resistance within weeds. Shortly after that, in 1957, wild carrots in Ontario, Canada, were discovered to have evolved resistance to a common herbicide (2,4 D). As new herbicides were developed and deployed, other weeds followed the course pioneered by the wild carrot. From 1954 to 1960 seventeen newly resistant plants were discovered *each year*. Beginning in 1996, the development of Roundup Ready crops allowed farmers to spray the effective herbicide glyphosate (trademarked as Roundup) directly on crops and kill only the weeds. This dependency simplified their approach to weed management and led to a reliance on glyphosate as the go-to herbicide. Since 1974, approximately nineteen billion pounds of glyphosate have been sprayed on the world's crops, with increasing concern for human and environmental health. For example, a 2019 study has uncovered the herbicide's role in increasing the flow of phosphorus from fields to waterways. Increasing phosphorous pollution in streams increases plant growth and eventual decomposition, which robs streams of life-giving oxygen. Of course, overreliance on a single chemical also made the evolutionary response easy for plants, and by the year 2000 glyphosate-resistant horseweed had evolved. In the next five years, several other glyphosate-resistant weeds were successfully competing with Roundup Ready crops. In 2017, at least thirty-five plants were resistant to glyphosate. Although the pace of evolution has slowed since the 1970s to an average of nine newly evolved resistances per year, the total number of herbicide-resistant weed populations is staggering. By 2017, there were more than five hundred species of resistant weeds representing workarounds to nearly every form of herbicide developed. The evolution of resistance is seen as a significant threat to global food security, because weeds cause the loss of one-third of the potential crop yield each year.

Much as plants have evolved resistance to herbicides, insects have developed resistance to pesticides and even crop rotations. The Colorado potato beetle may hold the record for outmaneuvering the chemical war waged against it. This yellow-orange arthropod sports ten bold stripes lengthwise on its hard, outer wing coverings, and though it is less than a half of an inch long, its appetite for potatoes is legendary. A native of North America, this pest first attacked potato crops in Nebraska in 1859 but now has spread throughout the United States, Asia, and Europe. The adult beetles and their larvae defoliate potato plants, and they do so with gusto. During their larval and adult life, each is capable of eating nearly eight square inches of a leaf (an area about forty times its body size). Excessive removal of leaves can cause complete crop failure. The potato beetle evolved resistance to DDT, dieldrin, and fifty-three other chemicals designed to limit its damage, including carbamates, organophosphates, organotins, acids, pyrethrins, isoflavins, neonicotinoids, lactones, spinosyns, and endotoxins produced by Bt (*Bacillus thuringiensis*), which are effective against many pests. This insect may be exceptionally well suited to detoxify poisons, because its natural hosts include members of the highly toxic nightshade family. How beetles overcome the effects of pesticides are varied and include enhanced metabolism, reduced absorption, and increased excretion of poisons. In addition to these physiological counterstrategies, some populations of beetles flee toxic areas and individual leaves, especially those treated with Bt.

The rapidly evolving Colorado potato beetle, here on a bittersweet nightshade. (Photo by Ryan R. Garrison)

Farmers around the world continue to fight Colorado potato beetles with applications of multichemical cocktails, changes in planting time, mulching, trenching, and encouragement of natural enemies. Not using pesticides in some fields reduces the evolution of resistance and favors non-resistant strains so that subsequent insecticide applications are effective. Even providing no-treatment refuges within fields can be useful because when the nonresistant beetles from sanctuaries mate with resistant beetles outside of them, they dilute the effect of genes conferring resistance and slow the evolution of super-resistance. Despite the lure of finding a chemical silver bullet in the fight against potato beetles, crop rotation appears to be a more effective strategy. Raising other crops that do not support beetles before planting potatoes again reduces the ability of pest populations to build up and evolve in response to pesticide use.

Nearly every farmer I talked with mentioned crop rotation as a strategy he or she used to fight pests, be they consumers of corn, wheat, or soy. However, evolution can foil even this strategy. Some populations of northern and western corn rootworms, for example, can counter the typical midwestern rotation of corn and soybean crops. And they do so by employing different strategies. Northern rootworms extend the length of time their eggs remain dormant in the soil, a period of suspended animation known as diapause, to two or more years. Typically, eggs laid in autumn spend the winter in the ground and hatch into larvae that feed on corn roots for a month and a half during the next summer. After a week or so of pupation, adult beetles emerge and feed on corn silk. By extending diapause across years, rootworm eggs wait out the year or two during which soybeans are grown instead of corn and hatch when corn roots fill the fields. In contrast, although larval and adult western rootworms also specialize on corn, they do not have extended periods of diapause. Instead, their parents lay eggs on soybean plants, so that the typical yearlong diapause results in larvae emerging as the next year's corn crop develops.

It is fair to say that the harder one tries to suppress a pest, the more likely it is to evolve countermeasures, be they chemical resistance, behavioral adjustments, or altered life histories. This response is merely the way that evolution by natural selection works: the stronger the force, the more rapid and extreme the evolutionary response. As long as the myriad traits enabling resistance are heritable—able to be passed on either genetically or culturally from parent to offspring—and pest populations are large enough to avoid random changes in their genetic composition or cultural traditions, then increased reproduction by individuals possessing resistant traits will transform the next generation into one with increased resistance. A few generations of such change produce the populations scientists have now documented as being resistant to every herbicide, insecticide, and antibiotic humans have developed. Evolution is truly awe-inspiring in its capacity to fit an organism to its changing environment, but this proficiency is also incredibly frustrating to farmers and doctors seeking cures for what ails us and our crops.

CHARLES DARWIN STARTED his treatise on natural selection by reminding readers of how human action has transformed domestic animals. Domestication occurs because artificial selection imposed by humans causes exceptionally rapid evolution. The domestic animals we house in agricultural settings also provide a new theater for the development of deadly disease organisms. As an international team of disease ecologists and public health investigators put it: "The intensification of agriculture in the twentieth century transformed the landscape for the infectious agents of farm animals." The results of this evolution are evident in diseases that plague swine and poultry and, increasingly, in disease outbreaks that move from farm to table. The emergence and spread of disease remind us that evolution is not merely of theoretical or economic importance; at times, it is a matter of our life and death.

Perhaps you remember the swine flu pandemic of 2009 or more recent, seemingly annual, eruptions of bird flu. As many as two hundred thousand people died from respiratory complications resulting from swine flu, most in Mexico, Argentina, and Brazil. Bird flu was mostly confined to poultry, resulting in the culling of millions of chickens, ducks, and turkeys in efforts to limit the spread of the disease to humans. Still, several hundred people died, mostly in China. Those were relatively mild outbreaks, considering that the similar Spanish flu of 1918 killed 2 to 5 percent of the entire human population (fifty to one hundred million people). Each of these flu viruses is a form of the influenza A virus that is continuously evolving within its hosts. The movements and interactions of wild birds, people, and our farm animals, especially pigs and poultry, stoke the flu's evolution.

The cells in the upper respiratory tract of pigs are naturally able to bind influenza viruses from both pigs and humans. In this way, the RNA of human and pig flu can end up in the same pig cell. (The characteristics of viruses are encoded in the structure of their RNA, which like DNA is composed of repeated subunits that encode both the identity and assembly order of specific amino acids, the building blocks of proteins.) As the pig and human viral RNA replicates, infected pig cells house vast numbers of each type of RNA, and because this RNA is not confined in a viral nucleus and is broken up into eight small segments, it is very prone to swapping bits and pieces during replication. Exchange even occurs between pig and human viral RNA. Frequent mutation and trading of genes creates new flu strains at an alarming rate, an estimated one million times faster in

Pigs serve as deadly mixing bowls for flu viruses.

viruses than in organisms such as ourselves, where mutation rates are low and merging of DNA segments from different species is virtually nonexistent. Rapid viral evolution within hogs produces strains that can bind with human cell membranes, strains that are increasingly virulent, and strains that are both. When a pig infected with the latter strain sneezes near a farmworker, pandemics, such as the Spanish flu, can result.

As we move farm animals and ourselves around the world and as birds migrate, we increase the potential for viruses to combine into new and deadly mixes. Influenza A viruses are widespread in migratory ducks, geese, swans, and shorebirds. As a result, rapidly evolving strains of flu are moved around the world seasonally as birds migrate. As humans, farm animals, and wild birds intermingle, contacting each other's waste or sharing a breath of contaminated air, they increase the potential to share viruses and evolve strains able to infect new hosts. In this way, variants of influenza A viruses once confined to wild birds can bind to and infect the cells of pigs and chickens, while those formerly limited to horses can infect dogs. We threaten our health and food security by stoking the flames of viral evolution.

Migratory waterfowl such as these Ross's geese naturally move flu viruses throughout the world.

Sometimes in our desire to control disease and to increase food production, we inadvertently play right into a virus's evolutionary gambit.

Humanity's collective appetite for chicken is staggering. Worldwide we eat about two hundred billion pounds of it annually, according to the Food and Agriculture Organization of the United Nations. Americans seem especially fond of the bird, eating more—an average of ninety-two pounds per person in 2017—than is consumed in any other country. Farmers have been able to keep pace with our increasing demand for poultry in large part by increasing the speed at which animals grow. The time it takes to produce a five-pound broiler chicken on an industrial farm, for example, has been halved (from ten to five weeks) since the 1950s. Shortening the lifespan of a broiler chicken allowed farmers worldwide to produce more than sixty billion birds in 2016. The United States is the worldwide leader in chicken farming, bringing nine billion broilers to market in 2017, for example, according to the National Chicken Council. Just the fact that we have a National Chicken Council ought to clue us in to the prominence of this bird!

The short lifespan of a broiler chicken not only quickly fills our larders but also directs the rapid evolution of chicken viruses in a novel way. Typically, the killing power, or virulence, of a disease that remains within a host for long periods is reduced to an intermediate level by natural selection. One theory posits that viruses are of intermediate virulence because those that are overly potent kill their hosts before they can infect others, while benign viruses produce few offspring and enable hosts to develop immunity. Throttling down virulence seems consistent with a pattern exhibited by the virus that produces Marek's disease in chickens. In response to the halving of broiler lifespans over the past sixty years, the losses from Marek's disease have also substantially increased. Increasingly virulent strains of the Marek's disease virus cause cancer and death in the chickens they infect. As losses have mounted, farmers have vaccinated their flocks, but immune strains have quickly evolved, and virulence has continued to climb. Losses of chickens to Marek's disease have become so severe that now nearly every broiler is vaccinated while still in the egg or within its first day of life.

It is likely that the rapid evolution of highly virulent strains of Marek's disease virus resulted from the use of vaccines as well as from the shortened lifespan of broilers. Vaccination prolonged the life of infected chickens, and while it controlled the spread of viral strains of low virulence, it afforded resistant and increasingly virulent strains a haven where they could survive and reproduce in a host kept alive by artificial means. In addition to this benefit, the typical cost of high virulence — killing the host and therefore shortening the period of disease transmission — was also relaxed because broilers lived just seven weeks before slaughter. The intensification of chicken production, while providing protein to a growing and increasingly affluent human population, has inadvertently increased the pathogenicity of a killer virus. Marek's disease hasn't yet jumped from chickens to other species, but just as it evolved increased strength, so too it may develop the ability to infect a broader range of hosts by bumping into a pig or human virus and swapping a bit of RNA. That possibility keeps public health officials up at night.

Intensive husbandry of farm animals affects our food security and health not only through the rapid evolution of novel viruses but also by stimulating the evolution of antibiotic-resistant bacteria. Most scientific research points to how the use of medically important antibiotics in both humans and animals contributes to the development of antimicrobial resistance in all sorts of bacteria. Bacteria causing staph infections, for example, now exist that are immune to treatment from penicillin, methicillin, tetracycline, erythromycin, and vancomycin. These resistances first started to become common in the 1960s and eventually prompted the development of a new class of antibiotics, oxazolidinones, to fight the disease in the 1990s. Staph evolved resistance to these more modern medicines in a decade.

Resistance evolves rapidly in bacteria because our drugs are strong selective agents that, while initially deadly to a large fraction of the bacteria causing an infection, favor the few individuals that are naturally resistant to treatment. Populations of resistant bacteria grow as resistant individuals quickly multiply and share their resistance genes, even with different bacterial species, by swapping bits of DNA, much as distinct viruses exchange bits of RNA. Obtaining a resistance gene by this horizontal gene transfer confers a nearly immediate advantage on its possessor. It is therefore essential to use these antibiotics only when medically necessary.

There has been debate over the degree to which agriculture spurs the evolution of antibiotic resistance, but after the World Health Organization in 2006 implicated animal husbandry in the development of drug-resistant microbes, the use of antibiotics on the farm has declined. As of 2018 the European Union, the United States, and many other countries now believe that there should be a ban or (in the United States) a phaseout of the use of medically important antimicrobials in food animals to enhance growth or improve feed efficiency. They further believe in bringing the applications of such drugs to treat, control, or prevent specific diseases under the oversight of licensed veterinarians. In the European Union antibiotic use for agriculture has steadily declined, as has antibiotic resistance of specific pathogens, while still allowing an increase in meat production. However, the level of antibiotic-resistant bacteria in the farms has not diminished to zero.

Despite governmental pledges to reduce the use of antibiotics on the farm, the threat of a new disease to crops or livestock can revive the careless use of important human medicines. In Florida, for example, a bacterial disease known as citrus greening threatens the wholesale destruction of an $8.6 billion industry. In response, orchardists convinced the Donald Trump administration to initiate the approval process in 2018 for the use of two antibiotics on their groves. Citrus growers hope that spraying hundreds of thousands of pounds of streptomycin and oxytetracycline on their fruits will arrest the deadly disease. Public health officials worry that the use of these medicines to treat diseases such as chlamydia and tuberculosis in humans will be compromised by the evolution of antibacterial resistance.

Bacteria are under continuous selection for resistance because antibiotics remain ubiquitous in our environment. Animals, including humans and livestock, excrete ingested antibiotics, which then end up in large quantities in our sewage, pastures, and feedlots. Sewage treatment is not wholly effective at degrading antibiotics, making treatment plants and livestock sewage lagoons or manure piles hotspots of antibiotic occurrence. These may be some of the most dangerous places on the planet, and they are a major reason many countries now limit the use of antibiotics in farm settings. There, bacteria can end up together in the guts of wild birds, such as ducks, gulls, and crows that visit farms and sewage plants, and it seems only then a matter of time before they swap DNA and gain superbug status: evolved resistance to multiple drugs.

The potential of some superbugs to evolve on the farm is especially frightening. The development of resistance to vancomycin is a case in point. Vancomycin is currently one of the standard treatments for patients suffering from infections caused by MRSA (methicillin-resistant *Staphylococcus aureus* bacteria). But vancomycin-resistant bacteria are common in a wide variety of settings. In the United States, vancomycin use in hospitals was frequent, and it was there that bacteria evolved resistance. These resistant bacteria then spread to urban and rural environments. In contrast, in Europe and elsewhere, resistance is thought to have originated on the farm because of widespread use of avoparcin, an antibiotic closely related to vancomycin, to enhance cattle health and growth. There, resistance spread from farm to urban and hospital environments. This difference was because avoparcin was banned in the United States but widely used elsewhere in feed for growth promotion. For example, Australia imported an average of

1,280 pounds of vancomycin per year from 1992 to 1996 for human clini-
cal use but over one hundred times more avoparcin (138,101 pounds) for
animal husbandry. What if there was a widespread transfer of vancomycin resistance to *Staphylococcus aureus* bacteria that were already resistant
to methicillin? Such a superbug could produce life-threatening infections
that would circumvent current treatments of MRSA. It only took a little
crow poop to convince me that the possibility of such evolution was far
from theoretical.

Professor Marilyn C. Roberts is an environmental health scientist at
the University of Washington. Marilyn and all the other microbiologists I
know are much more cautious than I am about contacting bird poop. She
diligently washes her produce, always wears gloves when handling birds
and samples, and regales me with tales of the hazards of my work and lifestyle. Her research has made her cautious; she merely knows all too well
the bacterial dangers that lurk on the farm and in wild animals. In 2013, we
worked together to collect samples from human sewage treatment plants,
dairy barns, and crows and test them for antibiotic-resistant bacteria. Our
findings surprised us both and got me in the habit of wearing gloves.

As Marilyn cultured the bacteria within crow dung, she discovered
that some were resistant to vancomycin and that every one of these bacteria
was simultaneously resistant to one or more other antibiotics. We found
crows harboring bacteria that had evolved resistance to clindamycin, erythromycin, minocycline, tetracycline, and chloramphenicol PLUS resistance
to vancomycin. These were super-duper bugs. And they not only had multi-
drug resistance but also were on the move. As crows travel between the
city and the farm, spending time at places where other multidrug-resistant
bacteria lurk, defecating where they find food or shelter, the chances that
vancomycin-resistant bacteria might share space with bacteria resistant
to methicillin increases. And so too does the possibility that multidrug-
resistant bacteria we can currently treat with antibiotics will evolve resistance, requiring new antibiotics that may not yet exist.

Mucking around in cow or crow poop is only one of the ways that
agricultural practices can increase the exposure of humans to rapidly evolving bacteria. We could ingest these bacteria by consuming contaminated

meat, dairy products, vegetables, or water. There are millions of such cases, often involving *Salmonella* or *Campylobacter,* each year in the United States alone. These bacteria are now able to move directly from human to human. In this way, MRSA from livestock is known to have spread to farmers, their family members, or other associates and those working on the farm. Fortunately, most animal bacteria rarely spread to humans. However, horizontal gene transfer can transfer the resistance genes they carry to human bacteria, including human pathogens. Thus, just as vancomycin-resistant bacteria moved from the farm through the community and into the hospital setting, other novel resistant bacteria too can jump from animals to people and cause challenging diseases.

The current understanding is that all use of antibiotics, whether for humans or livestock, contributes to the evolution of antibiotic-resistant genes and strains. These resistant bacteria can be maintained in the environment, agricultural setting, and human setting even after the antibiotic is no longer in use, making it difficult to eliminate them. Evolution of resistant bacteria in the hospital, home, farm, and overall environment has given humanity one of its most significant challenges in combating the rapid rise in bacterial resistance to some of our most essential medicines, with the result that treating common bacterial infections will become more difficult and costlier in the future.

THE EVOLUTION OF LIFE forms that share our farms fascinates me, but it also prompts consideration of how we might evolve our practices, both as farmers and as consumers, to enhance the well-being of the wild species that adapt to our agrarian ways.

THE STATISTICS CONCERNING THE EFFECT of agriculture on Earth are disturbing, and the paths toward human harmony with the land often seem impassable. The consensus among scientists in the social and natural sciences is that the current global food system drives climate change, biodiversity loss, change in land use, depletion of freshwater, and pollution of terrestrial and aquatic systems. I've read persuasive arguments detailing how we have already pushed a few of Earth's systems beyond the limits needed to maintain the environmental stability characteristic of the past ten thousand years. The current concentration of carbon dioxide in the atmosphere, rate of species extinction, and extraction of nitrogen from the atmosphere and its use as a fertilizer exceed such planetary boundaries. These boundaries are estimated values for key variables that are at safe distances from levels that profoundly disrupt Earth's biophysical processes. Crossing boundaries may cause dramatic shifts in systems, for example, hurricane cycles, with disastrous consequences to humanity.

In the face of such news, it is easy to lose hope, but I remain optimistic. Other colleagues share my optimism. Some remind us that major societal shifts occur in the face of unprecedented challenges. Adjustments in soil management following the Dust Bowl or in airport security after September 11, 2001, show us the power of adversity to promote change. Others point to the continuing emergence of new thoughts and innovative ways of living that overcome local and global constraints on sustaining Earth's resources. The proliferation of alternative energy sources and the many ways we've seen farmers and ranchers make room for wildlife on their land are recent examples. These instances will, I think, prompt societal change as we confront the grand challenge of feeding an increasingly large and affluent human population while also sustaining Earth's biological wonders. But although I'm hopeful, I am not patient. So, what can each of us do to effect change now?

Previous page: Bison can provide healthy and environmentally friendly red meat for our diet.

Though there are gaps in our knowledge and uncertainties in how Earth's systems respond to our actions, scientists and economists consistently point to three aspects of our food system that, if changed, would provide our growing population with the nourishment we need without compromising Earth's ability to sustain other life forms. I've introduced these adjustments, a Holy Trinity of sorts to guide future farming and eating habits: closing yield gaps with sustainable intensification, reducing food waste, and changing diets. If we adopt such changes, some argue, sufficient wildland would remain to slow the losses of plant and animals threatened by our agricultural enterprise. I agree that such changes will help provide what humanity needs in the future, but I am not confident in the claim that our care of nature also will be adequate. Instead, it seems to me that, in addition to adjusting our food system to spare some wildland and restore degraded land, we should also follow the motivational examples from the farmers and ranchers we've met in the preceding chapters to share our agricultural lands with our wild brethren better. Considering the trinity and the lessons from wildlife-friendly farmers, I find there is a lot we can do today to start making a difference. Pledging to do what we can and modeling our behavior to our families, friends, and peers will help create the societal shift necessary to meet the needs of people and the planet. Let's start with our diet.

I am not a vegetarian, but in researching this book, I have become much more aware of the environmental costs of my meat-based diet. Cattle and other ruminants, such as goats and sheep, release large amounts of methane, an uberpotent greenhouse gas, into the atmosphere as they burp and as their manure decomposes. And when the animals are fed a grain-based diet, their effect on the climate is exacerbated by poor digestibility and the conversion of land to supply feed. Growing grain for livestock takes a third of Earth's arable land off the table for wildlife. The dust from its harvest and ammonia released into the atmosphere from the fertilizers it demands shorten the lives of thousands of people each year. Improvements in feed, manure management, and growth rate may reduce these effects, but not below those inherent in other foodstuffs. These facts convinced me to treat beef and dairy products as luxury foods in my diet. I avoid grain-fed beef whenever possible and instead eat beef or bison raised on natural grasslands. When grazing on rangelands is consistent with the actions of native grazers, such as done by the Andersons and Milesnicks, then crops that could feed people are not subverted to feed animals or fuel our cars. This approach also uses less energy, water, and pesticides, all of which allow wildlife to coexist with livestock better.

Today's marketplace offers a wide variety of plant-based alternatives to meat. At my neighborhood store, I can choose between traditional tofu preparations as well as Tofurky, Smart Dogs, nut cheese, vegenaise, barbecued jackfruit, and Field Roast burgers, slices, and sausages. And then there are more radical substitutes, such as the Impossible Burger. This food is the plant patty that looks like a hamburger and even bleeds a bit when you take a bite! The blood comes from heme, an essential molecule that helps transport oxygen in animal blood. The heme in the Impossible Burger isn't from animal blood, but rather from soy. I was curious about this burger, so I enlisted my family to help with a taste test. Although the burger set us back a bit more than a typical gourmet sandwich in the United States (mine cost thirteen dollars), we all agreed it had the mouthfeel of beef and great taste. It was not the usual beany, stiff veggie burger. As its popularity and that of similar burgers increases and prices drop, it's sure to help many carnivores reduce their beef input. That can't happen soon enough for my fellow Americans. In 2018 we set a new annual meat-eating record, consuming an average of 222.2 pounds per person.

I am a long way from being entirely vegetarian, but I'm eating more grains, beans, nuts, seeds, vegetables, and fruits while limiting beef or bison to one meal a week. My wife and I aim to have a meatless dinner twice a week, and I'm rediscovering the pleasures of peanut butter and jelly sandwiches for lunch. We enjoy growing some of our food, a joy you can gain from a small plot of greens or a garden of fruits, vegetables, and a few livestock. In place of red meat, we eat wild-caught salmon, pork, or poultry two to three times a week and eggs once a week. Their environmental costs, which are not inconsequential, are much lower than those associated with cows, goats, or sheep. Chicken, pork, and large pelagic fish, such as salmon, also require less energy to produce or harvest and emit less greenhouse gas and pollutants, such as nitrogenous waste, than does the raising and harvest of alternatives, such as shrimp, carp, catfish, and tilapia. A 2018 assessment calculated that a global adoption of such a flexitarian diet that provides 2,100 to 2,200 calories per person per day while limiting red meat to one serving per week, white meat to half a portion a day, and dairy to a single portion per day would halve the projected increase in greenhouse gas emissions expected by 2050. Combining a flexitarian diet with the elimination of food-based biofuels (such as ethanol derived from corn) is estimated to allow the current crops we produce to feed about one billion more people without increasing cropland. And a diet that is low in red meat is also healthy!

Beyond what we eat, knowing where our food comes from is of increasing concern to consumers. As I visited farms and ranches, it became evident that to know how my food choices affect the land, I need to know more about my farmers. Do they tolerate native predators? Do they avoid groundwater depletion? Do they not till the ground after harvest so that soil vitality and organic matter are sustained? Are some of their lands fallow or permanently reserved? Do they maintain hedgerows or plant cover crops that promote native pollinators and pest-controlling birds and insects? Are their practices certified sustainable, organic, or in other ways? Do their products come from monocultures or a rich polyculture where productivity, resilience, and biodiversity are enhanced? Is there a strategic plan that minimizes their use of pesticides, herbicides, fertilizers, and water? Sometimes we can assess these attributes at the grocery store or restaurant. But there are more intimate ways.

Visiting the farm or shopping at farmers' markets or food hubs that aggregate, distribute, and sell local, source-identified products makes it easier to know your farmer. But even in these situations, you may not be able to discern farming practices that affect wildlife. I found it especially revealing to meet farmers and learn about their views concerning wildlife. These meetings started conversations that we both learned much from, and it got me to increase my purchases of local produce. Buying local—typically considered to be food raised within at most four hundred miles and often within a hundred miles of our homes—reduces the environmental cost of transporting food, assures that open spaces will exist near our towns and cities, and allows each of us to experience the costs and benefits of our agricultural production. The last consequence is especially necessary because it instills a sense of ownership in our actions. When our land and water is mismanaged, we feel the effects and are more likely to seek remedies.

Knowing farmers adds to my shopping experience. As I choose meats, I think about Andrew and Hilary guiding their herds as wolves and grizzlies go about their ways. Assessing veggies brings to mind Adam and his son laughing as they amend the fields at Oxbow while yellowthroats sing. Picking out a hot sauce, I think back on the rich birdlife around Frank's pond, grassy road edges, and old point rows. Perusing the bread and flour aisle takes me to the fallow fields Dean sets aside for water and larks at Wheat Montana. In deciding among bananas, I fret about Jenny's dad sheltering from pesticides in Costa Rica and always go for those with organic labels.

When we don't know the farmer, certifications can help guide us to environmentally responsible foods. However, much homework is needed to appreciate what a particular certification entails fully. In wine country, for example, sustainable certifications and organic certifications both let consumers know that vintners are stewarding their lands and reducing the environmental footprints of their businesses. One is not necessarily better than the other; each is more or less appropriate in a particular situation. And as we saw at the chocolate orchard in Costa Rica, certification may be economically infeasible for some small yet environmentally minded farmers. Most certifications, general organic labels for example, indirectly benefit wildlife, but some are unusually direct. Several bird-friendly labels are now available for coffee, beef, and maple syrup. Seeking out such products reinforces farm management for habitats used by sensitive birds.

Farm communities, and likely wildlife habitat as a consequence, benefit from fair trade practices. Industrial pineapple farms in Costa Rica owned by multinational subsidiary Pindeco have certified some of their fields as ISO Green and Fair Trade. This change by a company not always at the forefront of the worker or environmental concerns occurred in response to European consumer demands. As a result, pineapple laborers now have increased wages, enhanced social services, and better working conditions than they had a short time ago. Orlando Vargas, who has spent his life in the town of Volcán de Buenos Aires, worked in the pineapple fields thirty-five years ago, and after stooping for five hours a day to plant five hundred pineapple starts, he took home about thirty dollars a week. Today's workers make five times that amount and even earn overtime when working more than forty-eight hours a week. They also have better retirement benefits, housing, and discounted company stores. And the company gives back to their community to fund school projects and other needs. Orlando doesn't work for Pindeco, but he sees today's company as "helping the local economy and the well-being of nature and the local people."

As we eat healthier and demand more from our producers, we can also reduce food waste. Approximately a third of all the food produced by farmers and ranchers never reaches the market or our mouths. A 2018 assessment suggests that cutting this waste in half would reduce needed agricultural land by 740 million acres and decrease greenhouse gas emissions by one-fifth by 2050. If waste could be reduced by three-fourths, which is likely the maximum possible, we could lower environmental strain in the future by even more (overall from 9 to 24 percent lower than if we continue to waste food at the current rate). Such substantial changes require efficiencies at each step from farm to table. Let's start with our kitchen.

I've got an apple and two persimmons that are overly ripe. I'm planning to salvage them in a sauce, but some chefs are giving us the advice to go even further. Joel Gamoran, for example, shows those who watch his *Scraps* cooking show or seek help from his cookbook how to fully use leftovers and also make meals out of ordinary food waste, including fruit peelings and seafood shells. That isn't always practical in my kitchen, so I'm thinking carefully before buying perishable foods in great quantity and learning how best to preserve and share extras. When nothing else works, I'll return the spoils to the land by composting. Composting can extend beyond our kitchens. I saw it as a significant feature of many of the wildlife-friendly farms I visited. Inevitable waste—grape mash at California wineries, bread products at the dryland wheat farm, or rotten produce at small-scale organic farms from Costa Rica to Washington—is composted into nutrient-rich mulch that lowers the needs for fertilizer.

Composting food waste builds healthy soils.

Lowering the amount of food waste at restaurants and stores can magnify the savings we make at home. Efficient use of food by commercial kitchens saves waste and bolsters the businesses' bottom line. Simple measures can also help reduce waste by patrons. Small plate size has been shown to lower waste at buffet restaurants by 20 percent, for example. Leftovers from a variety of businesses and charities can be recycled to reduce waste and help those in need. Vending machines that offer excess food to the homeless, for instance, are cropping up in many cities. In developed economies, researchers suggest that encouraging the use of blemished or irregularly shaped produce and promoting sales of foods nearing best by dates can reduce store and household waste. The French grocery chain Intermarché provides a good model for such promotions.

Across a range of economies, scientists have argued that improved crop harvesting and storage, which in some locales includes greater access to refrigeration, can reduce waste. In developing economies, increases in transportation and market infrastructure can reduce spoilage from farm to market. Saving more of the harvest is one way that farmers can help meet future needs without increasing the land devoted to agriculture, but getting more from the area we farm is also essential.

Farms that fail to reach their full potential are said to have yield gaps. These lands produce fewer crops than they could mostly because of too little fertilizer or too little water. Globally, yield gaps run from 45 to 75 percent for most crops, but the factors limiting yields vary with the climate. Scientists broadly agree that such gaps can be greatly reduced, for example by increasing fertilizer used to grow wheat in eastern Europe and increasing water and fertilizer on Sub-Saharan African corn crops. And doing so would pay huge dividends for the planet. For example, relative to our current trajectory, we could by 2050 reduce agriculture's effect on the environment by anywhere from 3 to 30 percent by reducing yield gaps to 25 percent and lowering greenhouse gas emissions by improving rice farming, manure management, and feed additives that reduce cow burping. Reducing yield gaps to this degree would hinge on careful application of nitrogen and phosphorus fertilizers and better use of rainwater. If we cut yield gaps to 10 percent, increase the efficiency of fertilizers, abandon crop-based biofuels, and reduce greenhouse gas emissions primarily in rice and cattle farming, then we could lessen agriculture's impact on the environment by 11 to 54 percent by 2050.

Improved yields can be accomplished in traditional ways and further improved with new technologies. Dry ground can be fallowed to allow water reserves to recharge. Cover crops can be planted between crop rotations to enhance soil nitrogen. Manure and compost increase the fertility and moisture content of the soil. When improving degraded soils with such methods isn't feasible, then crafting plants that are better able to thrive in challenging conditions is also an option. Molecular biologists are busy revealing the full genetic complement of essential food crops and forage grasses so that they can be precisely altered to improve drought tolerance, nutrient use, and a host of other traits that stifle productivity. Such genetic modifications, while controversial, are sure to play an important role in future crop advances, as they did in fueling the past Green Revolution. And these will benefit the planet. However, despite the push to engineer the perfect plant, traditional breeding approaches are already producing

Leaving cover crops between rows of almond trees builds soil health and provides habitat for pollinators and other beneficial insects.

promising results. By selecting beans with extensive fine roots, for example, plant geneticists have developed a strain that is super-efficient at using soil phosphorus. These plants produce three times as many beans as usual and could be valuable in parts of Africa where soils degraded from a long history of agriculture would otherwise require extensive fertilizer. All of these old-school adjustments not only enhance the farmer's yields but also increase the suitability of the land for birds and other wildlife.

Technology provides myriad ways to close the yield gap and reduce the environmental degradation that is otherwise inherent in agricultural intensification. Delivering precise amounts of water, nitrogen, and phosphorus when and where they are needed by growing plants increases productivity without increasing polluting runoff. Optimal reallocation of nitrogen fertilizers from areas where they are overused to those where more is required could reduce pollution from runoff by 41 percent globally. Such optimization would require global cooperation, but accurate monitoring of water and nutrients in one's field can enhance productivity and minimize environmental degradation locally. Microsensors in fields, such as the ones advising Frank Muller when his peppers and tomatoes need watering, are increasingly common aids farmers use. Some, for example, obtain indirect measurements of plant health by monitoring a crop's near-infrared reflectance. Shifts in the reflectance from green to yellow indicate water- or nutrient-stressed plants, and overall drops in plant biomass indicate the presence of pests or weeds. In response, the farmer can treat specific areas in the field to bolster yield. Farmers who map their fields' variations in productivity can amend particular areas after the harvest to increase the next crop's yield.

New technologies are helping farmers find better uses for manure and wastewater that otherwise pose a hazard to the environment. In North Carolina, for example, Duke Energy is capturing methane from pig poop and using that to generate electricity for its clients. Its first such system is expected to produce 11,000 megawatt-hours per year, enough to power about a thousand homes. By 2021, the energy company anticipates filling 0.2 percent of all energy demands with agricultural methane. Some dairy farmers are taking the use of manure to even higher levels. Jeremy Visser tends a herd of two thousand cows on his farm outside of Stanwood, Washington. He has separated the fiber from wet manure and dried it for many years, producing soft, recycled bedding for his girls. Now he is working with state and tribal partners to capture the steam from the drying process with a Janicki Omni Processor that will filter and condense the steam into water suitable for human and herd consumption.

Technology has provided intelligent solutions to the twin problems of growing plants in nutrient-poor soils and limiting the runoff of fertilizers from farms. But when it comes to increasing crop yields by reducing pests and weeds, technological fixes rarely appear to be in harmony with the land. Effective pesticides have unintended consequences that constrain yields. Neonicotinoids, which kill pests that have yet to evolve resistance, also severely disrupt the pollination services that bees provide to farmers. Sulfoximine-based pesticides were introduced to replace neonicotinoids, but now this poison has also been found to limit bumblebee reproduction and colony growth. Rather than continuing to rely on human-made chemical pesticides, promising research suggests that harnessing natural microbes may be more effective. Farmers have used the bacterium *Bacillus thuringiensis,* which produces proteins that gut a variety of insect species, for decades. Many of the farmers I visited regularly use Bt. However, new microbes, such as the bacterium *Burkholderia,* are also proving effective at controlling a wide variety of pests from insects to nematodes. Powerful new abilities to edit the genomes of plants have other scientists excited about engineering plants to resist disease. In this way, rice has been genetically modified to resist bacterial leaf streak and blight.

Most experts agree that "technology alone won't save the farm." Instead, even with the new advances to control pests or increase plant resistance, many of the methods I saw farmers using remain important. These include crop rotation, crop diversification, planting native vegetation or installing nest boxes to bolster natural insect predators, flushing and mechanical weeding before planting, and spacing crops appropriately to either shade out weeds or dry out molds. The simple act of leaving difficult-to-farm areas natural is also an effective way to slow the evolution of pest resistance to nearby farming practices. Natural lands harbor pest enemies as well as pest populations that are not under the selective force of mechanical or chemical control. These pests slow the evolution of resistance by maintaining a diverse pest gene pool that includes nonresistant genes. Less traditional methods even rely on listening to communication among plants. Growing mint near soybeans, for example, increases the resistance of soy to tobacco cutworm larvae and a common rust fungus. It seems that the aromatic volatiles released by mint stimulate the soy plants to bolster their defenses, which include self-healing damaged leaves, producing foliage less palatable to pests, and attracting natural predators of the pests.

Changing how we grow, waste, and consume food can significantly reduce the projected future environmental effects of a larger and more affluent human population. Our total environmental impacts are expected to neither increase nor decrease by more than 15 percent relative to present values if we cut food waste in half, reap 75 percent of the productivity that current agricultural land can provide, and adopt a diet that provides 2,100 to 2,200 calories per day. This diet would be short of a flexitarian one; however, it would include no more than three quarter-pound servings of red meat per week, obtain less than 5 percent of its energy intake from sugar, and provide no more than five servings per day of fruits and vegetables. However, our environmental impacts in 2050 could be 20 to 55 percent *less* than current ones if we cut food waste by three-quarters, close yield gaps to 90 percent, and adopt a flexitarian diet that limits red meat to a single serving per week and white meat and dairy to a half and full portion per day, respectively. As optimistic as these projects are, there is a worrisome confounding issue.

Climate change will seriously challenge future agricultural production. Crops, livestock, and farming techniques will have to contend with increasing droughts, lengthened growing seasons, and unpredictability of extreme climate events. An extremely powerful winter hurricane, termed a bomb cyclone, pounded midwestern American farmers and ranchers in March 2019, giving them a scary preview of climate unpredictability. As the storm dumped heavy, warm rain onto a frozen and snowy land, epic flooding overwhelmed parts of Nebraska, South Dakota, Iowa, and Wisconsin. Four people drowned and hundreds were forced from their homes as roads and bridges were destroyed. Ranchers could not deliver hay and forage to their stock. In Nebraska, cattle losses were estimated to cost $400 million. Nebraskan farmers suffered an additional $440 million in crop damages. The Olters' farm I visited did not flood, but the soggy fields made work difficult in the spring of 2019. Sam shrugged off the challenge. He's seen wet springs before and quips that one thing about weather is "it's always changing." Past changes often juxtaposed productive years with destructive ones, but I fear that in the future most changes will be increasingly destructive.

As our actions destabilize climate and temperature and precipitation regimes shift, farmers will have to contend with changes in soil moisture, pests, weeds, and diseases. Our current reliance on intensively managed monocultures—60 percent of humanity's calories come from just three crops: wheat, rice, and corn—is a risky strategy in increasingly unpredictable times. Therefore, increasing the range of plants and domestic livestock, farming and ranching locations, and husbandry techniques would seem a wise way to hedge agriculture in an uncertain world. Indeed, this is no longer science fiction.

In the high, dry, and cold plains of Bolivia, farmers are facing unprecedented heat, drought, and outbreaks of new diseases and insect pests. In response, farmers are planting trees, growing new crops (especially disease- and drought-resistant varieties), improving soil and water management, shifting planting locations, changing cropping schedules, and increasing their reliance on livestock. When crops fail, farmers count on stress-tolerant wild relatives of commonly domesticated species, including quinoa and potato. Scientists researching this agroecosystem suggest that because of the myriad ways in which climate stress affects agriculture, crop portfolios must be kept diverse with a range of responses to the physical—temperature and humidity—as well as biological—diseases and pests—challenges expected in the future.

The challenges faced by Bolivian farmers is not unique. Around the world, farmers and ranchers are dealing with aspects of climate change, from unpredictable storms to unprecedented drought. Many governments react primarily by offering low-cost insurance against loss, but as our population grows, such loss will become increasingly intolerable. In the face of change and increased demand, we can all take a lesson from our Bolivian neighbors. In addition to adjustments in diet, yield, and waste, we will need to maintain local, native crop and livestock varieties as well as develop new strains that anticipate the many challenges of future conditions.

CHANGING OUR DIETS, wasting less, and growing more on land already farmed can meet the food demands of our expanding population, but how do we simultaneously ensure that other land is restored or maintained with nature in mind? As I talked with farmers and dug into the literature, it became apparent that part of the answer revolves around how governments incentivize farmers, how farmers benefit from maintaining biodiversity, and how society enables farmers to own or hold rights to the land. When we have a voice in government, we should raise it on behalf of farmers who seek to conserve wildlife and optimize crop production.

The march of corn and soy across the midwestern United States is in large part a response to government programs that have propped up markets for biofuels, beef, and high-fructose corn sugar. How a government subsidizes farmers and markets are myriad, including direct payments, production mandates, low-interest loans, cheap insurance, and tariffs and limits on imports. Nations around the world allocate billions of dollars to farm subsidies annually. The U.S. government, for example, spent more than $17 billion in 2015 to lower premiums farmers pay to insure crops and livestock, offset losses in disasters, such as storms and droughts, and directly pay for certain commodities. Similarly, the Chinese government uses substantial subsidies to balance grain farming between corn and soy. In the European Union, subsidies of nearly €59 billion in 2018, are increasingly tied to the needs of farmers by providing payments that bolster salary, stabilize markets, and enhance rural development.

While governments subsidize crop production, they provide less for conservation actions on farms. Undervaluing conservation is especially egregious on the farms I visited in the United States. In comparison to the $17 billion spent to advance crop production in 2015, the U.S. government provided farmers only $3.6 billion to reserve or restore sensitive wildlife areas. As I saw in Nebraska and Montana, such payments offer little incentive to farmers because they can earn much more by cropping rather than conserving the land. In Europe, subsidies for farmland conservation appear to have a more positive effect. There, Common Agricultural Policy reforms are helping to diversify cultivated areas by paying farmers to add perennial crops, such as olives, to lands that once raised monocultures of cereal grains. As a result, landscape diversity—and in response bird diversity—has increased in the once impoverished grain belt to the point that it equals the variety found in less intensive agricultural areas that spare natural lands and employ wildlife-friendly farming practices.

By supporting government investment in conservation rather than subsidies for unneeded crop production we can help slow the loss of biodiversity due to agricultural expansion in a powerful way. A 2017 survey of all 109 countries that signed the United Nations Convention on Biological Diversity in 1992 revealed that their investments in conservation from 1996 to 2008 reduced the loss of birds and mammals by a median of 29 percent per country. Investments were substantial—in total over $2.6 billion annually—and most effective at reducing biodiversity loss in low-income countries and those with many threatened species. The negative consequences of agricultural expansion were most severe in countries such as Malaysia and Peru that have little land devoted to agriculture at present and were most effectively held in check by countries with good governance (high quality of public service, independence from political pressure, and high quality of policy formulation and implementation). The authors of this study were optimistic in concluding that relatively small increases in conservation spending (one to five million dollars) by each country could further reduce endangerment and extinction of birds and mammals in the face of agricultural expansion. And these investments are lessened when we consider the full cost of crop production. The health costs of air pollution from corn production in the United States, for example, were estimated in 2019 at between fourteen and sixteen billion dollars. That equals almost two-thirds (62 percent) of the annual value of the crop. When we factor in such losses, conservation investments make economic sense.

Creation of environmental markets can also increase the worth of conservation actions on the farm. As society increasingly values a healthy environment, farmers may be able to sell environmental credits in nontraditional marketplaces. For example, farmers who reduce tillage to increase soil organic content or capture methane from manure could gain carbon credits that are purchased by industries and individuals interested in lowering their environmental footprint or meeting future standards. Setting aside wetlands and other habitats destroyed by development can earn credits in mitigation or conservation banks. Developers purchase these credits seeking to get variances, for example in building density. Environmental quality markets, such as the payments made to California rice farmers to flood rather than burn their fields, also pay farmers to reduce fertilizer runoff and silting of some streams. Similarly, reducing water use can earn credits in a water quantity market. Where solar, wind, or manure resources are abundant, farmers can generate energy and sell it to utilities that are mandated to seek renewable sources. Many of these markets are under the purview of government and industry, but some, such as those that conservation organizations create to pay farmers to delay harvest on behalf of tricolored blackbirds or bobolinks, can be bolstered by our donations. Participating in such efforts can help farmers conserve rare species that use their lands.

Many of the farms and ranches I visited attracted tourists to supplement their agricultural income. Visitors from around the world spend the night in the tiny houses and restored one-room schoolhouse on the Andersons' ranch. School groups, birdwatchers, and those wishing to learn about sustainable agriculture flock to the farms of Mastatal, Costa Rica. The income provided by agritourism allows these owners to remain on the land, improve their practices, and showcase their approach to farming. The benefits of ecotourism also spread beyond the boundaries of the farm to adjoining communities, increasing, for example, income at a small restaurant in Mastatal. In the landscapes I visited, inviting tourists to the farm mostly enabled land to be shared with wildlife. In the case of Hacienda Barú, the income from tourism fully replaced that from cattle ranching and permitted land to be spared from agriculture.

The imbalance between benefits to use farmland wisely versus intensively is a driving force behind the steady march of birds, such as the meadowlark, toward extinction. The diversity of farmers I met appreciate the birds and other wildlife that shared their fields. But their interest in conservation was tempered by economic reality. As a society, we should encourage the creation of policies that adequately reward farmers who devote some of their lands to nature protection. Favoring policies that directly link agricultural intensification to concurrent land sparing, for example, incentivizes improvements in food production while halting the spread of cropland into areas otherwise suitable for conservation. Allowing farmers to benefit from conservation stimulates wildlife-friendly strategies. Ranchers tolerate large carnivores when they get a premium for their beef. They control grazing along streams when they benefit from a vibrant fishery. Farmers are willing to lose some of their crops to monkeys, toucans, and coatimundis when tourists are eager to glimpse these beauties. The land is willingly left wild or enhanced when it harbors pollinators and predators that feast on a farmer's pests.

Gaining from conservation makes it easier for farmers and ranchers to share their land with wildlife, and this is a necessity in most parts of the world. As I considered the alternative strategies of sparing or sharing land with nature, it was obvious to me that although sparing may be possible in sparsely inhabited regions, it is less likely in areas already intensively farmed or settled. The costs of purchasing and restoring prime farmland often stifle reservation. On the other hand, sharing land with wildlife is always possible, it is inexpensive, and it may even repay the landowner with savings of energy, money, and time. Those I met who shared their land rarely did so for economic reasons. I was deeply moved when time after time a vintner, wheat farmer, rancher, or subsistence grower told me he or she worked their land with a gentle hand because "it was the right thing to do." Doing the right thing was personal to these agrarians because they had deep ties to the land. Their fathers and grandfathers showed them how to maintain the land's capacity for crops, which depended on its ecological integrity. To spoil the land would be disrespectful.

Building a sustainable bond with the land is most likely when farmers own or have rights to the property. The inability to purchase land remains a significant hurdle to the many young people who are interested in ecologically minded farming. One way around this obstacle was on display in the farms and ranches I visited in Montana and Washington. There, benefactors hired farmers and collaborated with them to establish a shared vision of land management and put it into action. Supporting government policies that provide low-cost loans, tax breaks, and other incentives to new farmers is another way we can all help get more young farmers on the land. As farmers age, creative incentive programs that match young would-be farmers with established farm owners become increasingly important as they help those seeking retirement to pass their heritage on to the next generation. In developing countries, many farmers are women without rights to the land. They have little incentive to farm with long-term sustainability in mind, especially if that comes at the cost of short-term productivity. Empowering women can also lower the need to intensify farming if we urge our policymakers also to increase women's access to family planning services, which research confirms lowers fertility rates and, as a consequence, population growth rates. Voicing our support for landscapes where social equity is high, traditional practices and human health are valued, and farmers have reliable access to local land can lead to improvements in conservation and food production.

AS I EXPLORED THE farms and ranches revealed throughout this book I reflected on the universal pride their owners held for their land. From small family estates to more substantial holdings, the farmers and ranchers I met were unanimously eager for me to view their properties, learn how they worked them today, and see how they maintained them for future yields. They freely shared their challenges and breakthroughs, their profits and costs, and their hopes for the next generation of agrarians that will work the land. They all aim to keep their holdings productive and able to support people and nature. To them, the farm is as Aldo Leopold wished—more than a commodity, indeed, a community.

The farmers I met were realistic, but as an ecologist in a pessimistic world, I was buoyed by their optimism. As we grow more food, waste less, and eat more responsibly, we may all become a bit more optimistic about the future of humanity and the planet. Our current rates of population growth and resource consumption are not sustainable, but neither is the path to a more viable lifestyle hidden. And our progress along the route will reflect our values as a society to current and future generations. We have a long way to go in shining up our tarnished reflection. As environmental philosopher Brian Henning noted, our industrial agrarian systems, "which divert 40 percent of all food to feed animals or create fuels, suggest that dietary and transportation preferences of wealthier individuals are considered more important than feeding undernourished people, or the stability of the wider biotic community." I suspect that few of us are comfortable being associated with those values. Fortunately, the wildlife-friendly farms and ranches I visited suggest a value set that is more palatable: it is possible to raise crops that provide healthy diets for people and meaningful employment in farming communities within the planetary boundaries necessary to sustain other life. Although the majority of today's farmland does not mirror this ethic, with our collective efforts, in the future, it can.

Notes

Preface

Berry (1977), Soule and Piper (1992), Salatin (2010), Shepard (2013), Fitzmaurice and Gareau (2016), and Yahya et al. (2017) discuss organic, ecological, and polyculture farming. Zimmerer (2010) and Hamilton et al. (2015) discuss agroecology and agroforestry.

Zimmerer (2010) offers a general framework for studying the response of biodiversity to agriculture and climate change.

Green et al. (2005), Hulme et al. (2013), Fischer et al. (2014), and Balmford et al. (2015) frame the land sparing versus sharing debate.

The *Washington Post* reports on the increasing number of young farmers in the United States (Dewey 2017). According to the National (USA) Center for Education Statistics, the number of college degrees awarded in agriculture and natural resources increased 41 percent from 2008–2009 to 2012–2014.

Aldo Leopold presented his land ethic in *A Sand County Almanac* (1949) and frequently discussed wildlife-friendly farming in his essays (1931, 1939a,b).

Chapter 1

The North American Breeding Bird Survey is described at https://www.pwrc.usgs.gov/bbs/.

Walk et al. (2011) analyze changes in the relative abundance of meadowlarks in Illinois.

The International Union for Conservation of Nature's Red List can be accessed at https://www.iucnredlist.org/.

Professor Faaborg is quoted from his introduction to Ornithological Monograph No. 64 (Askins et al. 2007); ornithologist Knopf is quoted in Knopf (1994).

Vickery and Herkert (1999), Conner et al. (2001), and the World Wildlife Fund (annual reports on project Plowprint available at https://www.worldwildlife.org/projects/plowprint-report) present statistics on the extent of grassland loss.

Vickery and Herkert (1999) and Askins et al. (2007) discuss agricultural intensification and loss of birds in the United States; Shrubb (2003), Reif et al. (2008), and Wilson et al. (2009) discuss the same for Europe.

Stevenson et al. (2013) estimate the influence of the Green Revolution on global land use. Foley et al. (2011) report on tropical forest loss to agriculture.

Barrett (2010) and FAO (2017) discuss world hunger despite a global food glut.

Balmford et al. (2015) is quoted concerning cultivating crops and keeping livestock.

Godfray et al. (2010), Tester and Langridge (2010), Foley et al. (2011), Tilman et al. (2011), Balmford et al. (2012), Robertson (2015), Tilman and Clark (2015), Erb et al. (2016), and Springmann et al. (2018) discuss the ecological consequences of feeding an expected population of nine billion in 2050 and eleven billion by 2100.

Seto and Ramankutty (2016) consider the importance of urban growth and agriculture in calculations of future land use.

Creutzig (2017) estimates the needs exceeding Earth's capacity.

Pretty et al. (2006), Tester and Langridge (2010), Godfray et al. (2010), Mueller et al. (2012), and Mueller and Binder (2015) discuss yield gaps.

Tilman et al. (2002), Pretty et al. (2006), Baulcombe (2010), Gebbers and Adamchuk (2010), Tilman et al. (2011), and Eisler et al. (2014) discuss sustainable intensification and resource conservation agriculture, including improved livestock management.

Godfray et al. (2010), Foley (2011), Foley et al. (2011), Tilman et al. (2011), Tilman and Clark (2014, 2015), Jahn et al. (2015), and Erb et al. (2016) discuss reducing waste and adopting a more plant-based diet.

FAOSTAT (http://www.fao.org/faostat/en/#data/QC) provides counts of livestock worldwide, and Koneswaran and Nierenberg (2008), Ripple et al. (2014), and Hilborn et al. (2018) analyze livestock emission of greenhouse gasses.

"Growing agriculture sustainably while shrinking its footprint dramatically" is quoted from Foley et al. (2011).

Crist et al. (2017) discuss decelerating population growth as a path to sustainable agriculture.

Walk et al. (2011) note that the "meadowlark's pattern of remaining widespread at a low-density rather than becoming localized has probably kept many people from realizing how severely their abundance has declined."

Chapter 2

Diamond (1993), Wood (2010), Guzmán and Weisdorf (2011), Henn et al. (2012), McGowan (2016), Tucci and Akey (2016), Stringer and Galway-Witham (2017), and Siska et al. (2017) discuss the early evolution of humans, the Neolithic Revolution, and the spread of people across the world.

Varki and Altheide (2005), Prüfer et al. (2012), and Manuel et al. (2016) present our genetic similarity with chimps and bonobos.

Diamond (1993), Mannion (1999), Patin et al. (2009), Zeder (2011), Dow and Reed (2011), McDermott (2016), Arendt et al. (2016), and Lazaridis et al. (2016) recount the early rise of agriculture.

Mannion (1999), Zeder (2006), Carneiro et al. (2011), Meyer and Purugganan (2013), Larson and Fuller (2014), Wilkins et al. (2014), Agnvall et al. (2015), Arendt et al. (2016), Zheng et al. (2016), Caspermeyer (2017), Choi et al. (2017), and Loog et al. (2017) investigate domestication of plants and animals.

There are two schools of thought concerning the domestication of dogs from wolves. My reading and discussion with wolf experts lead me to favor the idea that wolves initiated the process as I present in the text. However, others believe that early people started the process by raising wolf pups, which were increasingly bred for their compatibility, cooperation, and usefulness to people. Current studies of cooperation of wolves and dogs with familiar people suggest that wolves quickly initiate cooperative behavior (Range et al. 2019). Also, although wolf pups readily associate with humans that rear them, their attachment is much less than is seen between humans and dog puppies (Topál et al. 2005). These findings suggest that wolves would have been capable of starting the domestication process and that further selection and breeding by humans was also involved.

Smith et al. (2015a,b) discuss the spread of wheat into Britain.

Richerson et al. (2001), Pearce-Duvet (2006), Richerson et al. (2010), Polley et al. (2015), Scott (2017), and Kohler et al. (2017) discuss the challenges of agrarian life, including disease, general hardship, coercion, and selection for novel genes and cultures.

Farrell et al. (2001), Mueller et al. (2005), and Chomicki and Renner (2016) review the evolution of agriculture in insects and other animals. A pictorial review is also available from the BBC at http://www.bbc.com/earth/story/20150105 -animals-that-grow-their-own-food.

Koneswaran and Nierenberg (2008), Tester and Langridge (2010), Godfray et al. (2010), Foley et al. (2011), Tilman et al. (2011), Balmford et al. (2012), Ripple et al. (2014), Robertson (2015), Tilman and Clark (2015), Ellis et al. (2016), Erb et al. (2016), Maxwell et al. (2016), Springmann et al. (2018), and Hilborn et al. (2018) review the influence of agriculture on Earth's ecology.

Updated organic farming statistics are from the USDA's National Agricultural Data Service's 2014 and 2015 Organic Certifier Data and the 2012 U.S. Census of Agriculture (table 42, Organic Agriculture).

Leopold (1931, 1939a,b) extolled wildlife-friendly farming practices that have been updated by Soule and Piper (1992), Green et al. (2005), White (2008), Wilson et al. (2009), Quinn et al. (2012), Wright et al. (2012), Gabriel et al. (2013), and Teillard et al. (2015).

Green et al. (2005) report on payments by the EU for wildlife-friendly farming.

Green et al. (2005), Hulme et al. (2013), Fischer et al. (2014), and Balmford et al. (2015) reveal the debate over land sparing versus land sharing.

Shepard (2013) discusses the possible yields from a polyculture.

Erb et al. (2016) evaluate the ability of a variety of agricultural schemes to feed the world in 2050.

Godfray et al. (2010) and FAO (2017) provide statistics on food waste.

Crist et al. (2007) discuss societal changes needed to lower population growth.

Chapter 3

The farmers and former residents of Ong, Nebraska, preferred that I not use their actual names in recounting my interviews, so I have used fictitious ones in this chapter.

Dr. Seuss's *The Lorax* was published in 1971 by Random House Books for Young Readers.

EWG Resources (https://farm.ewg.org/) provides statistics on federal subsidies from 1995 to 2014.

Childs (2012) spent time inside a cornfield, recording the few critters that lived there.

Crosby et al. (2015) document losses of bobwhite quail and other grassland birds in the United States.

You can obtain statistics from a variety of U.S. North American Breeding Bird Survey Routes, such as I did for Nebraska's forty-six routes at https://www .pwrc.usgs.gov/bbs/.

I obtained values (in acres) of cropland, farmland, corn and other crops, weed con-trol, insect control, fertilizer, pasture, and harvesting methods in Nebraska from 1870 to 2012 from the USDA's census of agriculture at https://www .nass.usda.gov/AgCensus/. I obtained values for the Palmer drought index from the National Centers for Environmental Information at https://www .ncdc.noaa.gov/temp-and-precip/drought/historical-palmers/.

I consulted Vickery and Herkert (1999), Askins et al. (2007), MacDonald (2012), Harrison et al. (2017), and Niemuth et al. (2017) for reviews of the plight of grassland birds.

My colleagues and I report on the relations among weather, habitat change, and suburban birds in Shryock et al. (2017).

General information on black-tailed prairie dogs and their conservation status can be found at https://www.fws.gov/mountain-prairie/es/blackTailedPrairie

Dog.php. The response of birds and vegetation to their presence is from Lipinski et al. (2014).

Merrill et al. (1999), Svedarsky et al. (2000), Niemuth (2005), and Harrison et al. (2017) investigate the response of greater prairie chickens to land use.

I obtained statistics about the CRP program from the USDA's Farm Service Agency (FSA) at https://www.fsa.usda.gov/programs-and-services/conservation -programs/conservation-reserve-program/index. Specific allocations of CRP land in Nebraska are from EWG's conservation database, https://conser vation.ewg.org/crp.php?fips=31000&regionname=Nebraska, and the rental rates per acre are from the January 2017 *Newsletter* from the Nebraska State FSA Office. Allocation of federal funds to support CRP rental is provided by the 2014 Farm Bill (http://www.thefarmbill.com/). The Congressional Budget Office's estimation of costs for the 2018 Farm Bill can be found at https:// crsreports.congress.gov/product/pdf/R/R45525.

Herkert (2009) investigates the relation between CRP establishment and bird trends.

Wuthnow (2011) discusses the social fabric of Middle America.

The USDA National Agricultural Statistics Service and the annual census of agriculture (https://www.nass.usda.gov/AgCensus/) report farm numbers and size.

Thurin (1986) describes Ong, Nebraska.

Hill et al. (2019) estimate the health costs, including premature mortality, of corn farming to the population of the United States.

Leopold's suggestions for keeping farmland birds common appear in many of his essays (e.g., 1931, 1939a,b).

Allocations in the 2014 bill (http://www.thefarmbill.com/) and changes through time are from https://www.fsa.usda.gov/programs-and-services/conservation -programs/crp-grasslands/index. Details for the 2018 Farm Bill are from https://crsreports.congress.gov/product/pdf/R/R45525.

MacDonald (2012) studies the responses of prairie birds to field strips.

Wright and Wimberly (2013) quantify the recent expansion of the U.S. corn belt.

The USDA Economic Research Service provides statistics on the use of corn and other crops in the United States, which can be found at https://www.ers .usda.gov/.

Chandler et al. (2013) link land sparing to intensive coffee production.

Foley (2013) discusses the use of corn in the United States and solutions to shift to a more sustainable pathway.

Leopold's quotation about corn is from his 1947 foreword to *Great Possessions,* the working title of *A Sand County Almanac.* This unpublished foreword is discussed and reprinted in Ribbens (1987).

Chapter 4

Giordano (2009) reviews groundwater issues. Gleeson et al. (2012) measure groundwater depletion worldwide, and Dalin et al. (2017) tie it to specific crops and international trade of food products.

Granatstein (1992) provides a review of dryland farming in the intermountain West. I quote this report's comment "extolling the virtues of fertile land and nutritious crops."

You can learn more about Wheat Montana Farms at http://www.wheatmontana .com/.

Wheat Montana is in the Guinness Book of World Records under "Fastest field to loaf—microwave" (www.guinnessworldrecords.com).

Beason (1995) reviews the natural history of horned larks. Their abundance in agricultural settings in Illinois is surveyed by Walk et al. (2011).

I used the Breeding Bird Survey (https://www.pwrc.usgs.gov/bbs/) to calculate horned lark and California gull trends in Montana from 1968 to 2015.

Herkert (2009) reports the effects of the CRP program on birds.

Winkler (1996) reviews the natural and cultural history of California gulls.

Scientists from Montana Audubon and Montana Fish, Wildlife and Parks surveyed colonial-nesting water birds in the state (Wightman et al. 2011).

The domestication of wheat is discussed by Mannion (1999) and Meyer and Purugganan (2016) and summarized in Wikipedia, s.v. "wheat."

I queried the worldwide production of wheat using FAOSTAT (http://www.fao .org/faostat/en/#data/QC).

Statistics for the city of Three Forks, Montana, are from http://www.city-data.com /city/Three-Forks-Montana.html#b.

Chapter 5

Marti et al. (2005) review the natural history of barn owls.

Visitor statistics for the Napa Valley are from https://www.visitnapavalley.com.

I recommend Tattersall and DeSalle (2015) for an introduction to the natural and cultural history of wine, including diseases and the influence of land on a wine's character.

Matt Johnson's initial owl research is published in Browning et al. (2016) and Wendt and Johnson (2017).

Xeronimo Castañeda's M.Sc. thesis was completed in 2018 at Humboldt State University.

Glausiusz (2018) reports on the barn owl programs in the Middle East.

Insect consumption by western bluebirds is reported in Guinan et al. (2008).

Jedlicka et al. (2011) and Howard and Johnson (2014) use experiments to investigate the role of bluebirds in vineyards.

Information on certification programs in California's vineyards can be found at www.napagreen.org, www.fishfriendlyfarming.org, and www.ccof.org.

The organization of next-generation vintners in California is described at www .ngwine.com.

Sara Kross's research on barn owls and falcons in California vineyards appears in Kross et al. (2016), and her work on falcons in New Zealand is reported in her dissertation (2012) and in a series of publications (Kross et al. 2011, 2012, 2013).

Professor Kremen's research on hedgerows, pollinators, and other beneficial insects includes Kremen et al. (2012), Morandin et al. (2014, 2016), and Karp et al. (2015). A general synthesis of these findings and those from around the world are reported in Lichtenberg et al. (2017). Additional research into the pest-removal services provided by birds on farms can be found in Kellerman et al. (2008), Railsback and Johnson (2014), and Milligan et al. (2016).

Sara Kross's research on the influence of hedgerows on birds and the insects they eat includes Kross et al. (2016) and Heath et al. (2017).

General considerations of diversified farming systems and wildlife-friendly farming including the land sharing and land sparing debate in highly intensive agricultural landscapes include a back-and-forth exchange between researchers published in 2011 in *Science* (334:593–595) and Kremen et al. (2012) and Garfinkel and Johnson (2015).

Chapter 6

You can learn about Oxbow Farm and Conservation Center at www.oxbow.org.

If you ever find a banded bird in North America, please report it to the Bird Banding Lab, https://www.usgs.gov/centers/pwrc/science/bird-banding -laboratory?qt-science_center_objects=0#qt-science_center_objects.

I summarize my research in the suburbs of western Washington in *Welcome to Suburdia* (Marzluff 2014).

Guzy and Ritchison (1999) review the natural history of common yellowthroats; Ritchison (1991) describes their song.

Shrubb (2003) emphasizes the importance of farms to wintering birds.

You can learn more about the use of flooded rice fields by birds at Cosumnes River Preserve and throughout the Sacramento Valley at www.cosumnes.org; at https://www.audubon.org/news/sacramento-valley-rice-farmers-flock-save

-birds; and from Matt Jenkins's exposé "On the Wing," in the *Nature Conservancy Magazine,* August–September 2014.

The e-bird program is detailed at www.ebird.org.

Johnson and Igl (1995) measure the benefits of CRP lands to yellowthroats.

Montgomery (2017) celebrates the importance of building healthy soils on farms, no-till methods, regenerative agriculture, and the various experiments at the Rodale Institute.

Goulson (2015), Woodcock et al. (2016), Cressey (2018), and Butler (2018) discuss the plight of bees and the crucial role of neonicotinoid pesticides in pollinator declines. Kaiser-Bunbury et al. (2017), Carvell et al. (2017), Winfree et al. (2018), and Hass et al. (2018) review the importance of intact and restored ecosystems to pollinators.

Cutright (1989) reprints the responses of Meriwether Lewis, John Townsend, and Elliott Coues to camas root.

Zumkehr and Campbell (2015) discuss the potential of foodsheds in the United States to supply local demands.

Lebel et al. (2008) study Thai shrimp farmers and the sustainability of their farms. Cooke et al. (2008) introduce the global governance of food by examining trade among a complex web of actors.

I quoted the mission of Oxbow Farm & Conservation Center from its website, www.oxbow.org.

More about the International Farm Transition Network can be found at www.farm transition.org.

Cui et al. (2018) report on the adoption of sustainable practices by millions of Chinese farmers.

Chapter 7

More about Rancho Mastatal can be found at ranchomastatal.com. Finca Seimpre Verde is described at fincamastatal.com.

Coulson et al. (2018) document the relationship between caracaras and tapirs.

Bowers and Bailey (2001) review the history of cacao farming and diseases that have reduced productivity. Rice and Greenberg (2000) discuss the sustainability of small-scale cacao cultivation. More information about La Iguana Chocolate can be found at www.laiguanachocolate.com.

The ecolodge at Villas Mastatal is described at villasmastatalcr.com/en.

I obtained statistics on Costa Rican exports and land conversion from the CIA World Factbook (www.cia.gov/library/publications/the-world-factbook/), FAOSTAT (http://www.fao.org/faostat/en/#data/QC), and Monagabay.com.

Komar (2006) reviews the biological benefits of shade-grown coffee, and Blackman and Naranjo (2012) and Chandler et al. (2013) investigate the environmental effects of its certification in Costa Rica.

The criteria for certifying a coffee farm as bird friendly by the Smithsonian Institution can be viewed at https://nationalzoo.si.edu/migratory-birds/bird-friendly-farm-criteria.

Zuckerman's comment about palm oil is quoted from her 2016 article. Furumo and Aide (2017) characterize the land cleared for oil palm in Latin America. Literature concerning the effects on bird and mammal diversity of oil palm plantations includes Azhar et al. (2013), Yue et al. (2015), Vijay et al. (2016), and Yan (2017). Yahya et al. (2017) report the positive response of birds to diverse small farms within a monoculture of oil palm.

Perfecto and Vandermeer (2010) discuss land sharing versus land sparing in the neotropics, arguing for the value that a diverse landscape with small-scale, wildlife-friendly farms offers biodiversity.

Frishkoff et al. (2014) document the response of avian evolutionary history to intensive and low-intensity farming in Costa Rica. Professor Daily and her co-workers are quoted in this article.

Leck (1979) documents extinctions in an Ecuador forest preserve.

On corporate farms in Costa Rica, as elsewhere in the tropics, greed and lack of regulation often exposes farmworkers and those who live nearby to dangerous chemicals and difficult working conditions. Lawsuits claiming increased sterility, cancer, and birth defects by banana plantation workers exposed to insecticides, such as Nemagon, are still pending in Costa Rica's courts. Some exposure comes from handling, even wearing as raincoats, the blue polyethylene bags used to cover developing bananas. The bags increase yield and fruit quality by raising temperature, reducing UV rays, and limiting fruit damage during high winds. They often contain insecticides, such as chlorpyrifos, which has been linked to developmental defects in mice and children. Though still used widely in agriculture, its application is limited in residential settings and on some crops in the United States. Other workers contact toxins during accidental soakings by crop dusters applying herbicides and pesticides.

De Ponti et al. (2012), Ponisio et al. (2015), and Crowder and Reganold (2015) review the yield and profitability of organic versus conventional agriculture.

Chapter 8

The biological diversity of the Osa Peninsula is discussed on the Osa Conservation webpage, http://osaconservation.org/about-the-osa-peninsula/. You can read more about the Piro Biological Station and Osa Verde farm at http://osaconservation.org/visit-the-osa/research-stations/.

The connection between Cyclone Enawo and vanilla prices is detailed by Pam
Wright for weather.com at https://weather.com/news/news/vanilla-prices
-cyclone-enawo-madagascar.

Pinaria et al. (2010) discuss the effect of *Fusarium* fungi on vanilla production.

Leopold (1939b) suggested that the farm's landscape is its owner's self-portrait.

Whitworth et al. (2018) survey curassows and tinamous in a variety of forest types
at Piro, where hunting is not allowed.

Watson and Venter (2017) describe the "half-Earth" initiative.

Hanspach et al. (2017) investigate the ways in which biodiversity conservation and
food security interact in the global South.

Chapter 9

I learned about Montana's farm and ranch industry from the Montana Department
of Agriculture, https://agr.mt.gov/AgFacts.

The culling of wildlife by the U.S. government's Wildlife Services is published
annually at https://www.aphis.usda.gov/aphis/ourfocus/wildlifedamage/SA
_Reports.

Ripple et al. (2014) estimated greenhouse gas emissions by cattle.

Mueller (1999) describes the vocalizations of Wilson's snipe.

Niemeyer (2010) and Clark and Rutherford (2014) describe strategies, such as the
use of fladry and range riding, to coexist with wolves and other predators.
Treves et al. (2016) review efforts to test the efficacy of such strategies.

Savory (1988) describes holistic resource management.

Bud Williams describes Low-Stress Livestock Management at www.stockmanship
.com.

You can learn more about Yellowstone Grassfed Beef at https://www.yellowstone
grassfedbeef.com.

The status of the grizzly population in the lower United States is updated at https://
www.fws.gov/mountain-prairie/es/grizzlyBear.php.

Chapter 10

Díaz-Pascacio et al. (2018), Malan et al. (2018), and Miller et al. (2018) discuss the
influence of cattle on streams.

Hausen (2013) chronicles the Milesnicks' award.

The short-term rotational grazing that the Milesnicks and Andersons employ is
based on Savory (1988).

Vernal pool biology and conservation is presented in the U.S. Fish and Wildlife
Service (USFWS), Region 1, recovery plan (2015).

Use of grazing for conservation of wetlands is researched in Alberta by Miller et al. (2018), in Oregon by Charnley et al. (2018), and along the Wadden Sea by Mandema et al. (2015).

You can learn more about Knaggs Ranch and how they raise young salmon in flooded rice fields at https://caltrout.org/tag/knaggs-ranch/.

Renfrew et al. (2015) describe bobolinks and their natural history. Mather and Robertson (1992) discuss their flight displays.

The North American Breeding Bird Survey can be queried at https://www.pwrc .usgs.gov/bbs/.

Walk et al. (2011) speculate on the extinction of bobolinks in Illinois.

Perlut et al. (2006), Perlut and Strong (2011), and Renfrew et al. (2017) discuss the challenges of hayfield harvest and South American grain fields to bobolinks.

You can learn more about the Bobolink Project at https://www.bobolinkproject .com.

Beedy et al. (2018) describe the natural history of tri-colored blackbirds. The work of Audubon California to conserve this species is described at ca.audubon .org/birds-0/tricolored-blackbird.

Tucker (2013) reports on the loss of the MZ Bar Ranch by the Milesnicks.

You can learn more about bison management in Montana at https://www.nps.gov /yell/learn/management/bison-management.htm.

Allan et al. (2017) discuss the integration of cattle and wildlife in Kenya.

Chapter 11

You can read more about Jack Ewing and Hacienda Barú in his books (Ewing 2005, 2015) and at www.haciendabaru.com.

De Camino et al. (2000) report forest cover statistics for Costa Rica.

The path of the tapir biological corridor is described at www.asanacr.org.

Blasiak (2011) reports on President Chinchilla's 2011 comments and discusses the ways in which the Costa Rican government increased the nation's forest cover.

Walter Rosen coined "biodiversity" in 1985.

Hoelle (2012) studies the cattle ranchers of Amazonia.

Betts et al. (2017) review the influence of forest loss on biodiversity.

Chapter 12

Marzluff and Restani (2018) recount the extinction of birds, including Carolina parakeets.

I counted the number of extinctions and their cause using www.redlist.org.

Frishkoff et al. (2014) study the loss of avian evolutionary history in response to agriculture in Costa Rica. Dross et al. (2017) conduct a similar study in France. Morelli (2018) relates bird diversity and evolutionary history to agriculture in Italy.

Palumbi (2001) makes the case for humans being the greatest evolutionary force by reviewing how our actions affect the evolution of other species. Marzluff (2014) does so similarly for bird responses to urbanization.

Darawshi et al. (2017) study adaptation in short-toed snake-eagles.

Buij et al. (2013) relate grasshopper buzzard diets to land use.

Birds adjust their migratory routes to the vagaries of agriculture, including those of whooping cranes (Teitelbaum et al. 2016), sandhill cranes (Lacy et al. 2015), and crows (Marzluff and Angell 2005).

Mason and Unitt (2018) measure the change in horned lark coloration in response to agriculture.

Sætre et al. (2012) and Riyahi et al. (2013) study the evolution of house sparrows.

Marzluff and Angell (2005) propose the coevolution between human and crow cultures, including that involving avoidance of guarded crops. Swift and Marzluff (2015) research the response of crows to dead conspecifics. Clucas et al. (2013) measure the flight distance of crows in response to approaching humans.

Turcotte et al. (2017) review the general influence of farming on the evolution of other species.

Thomas (2015) documents increased plant speciation in response to agriculture.

Vigueira et al. (2013) and Turcotte et al. (2017) discuss weedy rye, which evolved its unique adaptations in 60 generations, and the rapid evolution in barnyardgrass.

Hébert et al. (2019) report on the link between glyphosate overuse and phosphorous runoff into agricultural watersheds.

Palumbi (2001), Délye et al. (2013), Vigueira et al. (2013), and Shaner (2014) chronicle the evolution of herbicide and pesticide resistance in weeds.

Turcotte et al. (2017) describe pesticide resistance in Colorado potato beetles and the evolutionary responses of rootworms to various crop rotations.

Atkins et al. (2012) provide the quotation about intensification of agriculture transforming infectious diseases of farm animals.

Simonsen et al. (2013) investigate the 2009 influenza pandemic and recount the history of other flu outbreaks.

Morgan (2006) discusses the evolution of influenza in agricultural settings and details the ways in which viruses invade hosts and replicate. RNA, which like DNA is a large molecule composed of repeated subunits of sugars, phosphates, and bases, consists of a single strand rather than the two strands that wind together into the familiar double helix of DNA. Despite their slight differences, the sequence of bases in both RNA and DNA encode both the identity and assembly order of specific amino acids, the building blocks of proteins. The proteins synthesized by reading the DNA or RNA code control the crucial functions within a cell. Slight changes in the RNA and DNA code (such as might arise frequently when new copies of a cell's RNA are made) may change the resulting proteins and therefore an organism's inward workings and outward appearance. For example, a slight mutation in the hemoglobin protein synthesized by DNA reduces its ability to carry oxygen in the blood and produces the condition we know as sickle cell anemia. This condition is inherited as offspring obtain this slightly changed (and here detrimental) form of DNA from their parents.

Viruses infect animals by binding to the outer membranes of their hosts' cells and either directly penetrating the membrane or inciting the cell membrane to fold inward, encircle the germ, and pinch off a pathogen-laden capsule through a process called endocytosis. Both routes require that the virus possess a protein that is accepted by receptor sites on the host's cell membrane. Specificity of the binding site of the virus and receptor site of the host normally act as a key and lock to limit infection to hosts that have coevolved to some extent with their natural viruses. Once inside the cell, the virus spews its RNA into the host cell, and the viral genome begins to replicate—its form of reproduction. Viruses do not contain their RNA in a nucleus like we do, so as the virus reproduces, RNA builds up inside the host cell. During this frenzied growth phase in the viral infection, mutations are common, and the resulting population of viral RNA becomes quite variable. As the virus multiplies and diversifies within its regular host, it is rarely fatal. To do so would limit the virus's ultimate reach. From the virus's point of view, it is better to sicken its victim just enough to assure widespread dispersal beyond the host, for example by provoking bouts of coughing, sneezing, or diarrhea.

Viruses evolve at a super-quick pace for two main reasons. First, the RNA that houses the genetic code within a virus mutates faster than does our DNA. This hypermutability occurs because DNA is replicated within the confines of our cells' nuclei (a process that proceeds the formation of new cells, including sperm and eggs) and the enzymes controlling the process frequently check their work and correct their mistakes, assuring transfer of a high-fidelity copy of the parental DNA to each new cell. RNA does not have such proofreading capacity, so errors in transcription (a form of mutation) accumulate in the hosts' cells. Second, in some cases the RNA from viruses usually infecting different beasts can mix. On the farm, pigs often host this dangerous genetic swap meet.

When viral strains from different hosts mix, new combinations can form that decouple the balance between severity and mobility that evolves within a long-standing host-virus duo. Although the separate, species-specific strains are mostly benign to their hosts, their combination produces deadly results to at least some species; swine flu and bird flu kill few pigs and chickens directly (countries may require culling of flocks and herds to prevent the spread to people), but they can be deadly to humans. The swine flu (designated H1N1 to represent the form of hemagglutinin and neuraminidase found in the virus) is an example of one such deadly concoction involving bits of virus from pigs, humans, and birds. Bird flu (H5N1) is a common pathogen of wild birds that is not yet a mix of viral genes from different species.

You can learn more about the National Chicken Council at https://www.national chickencouncil.org.

Atkins et al. (2012) investigate the evolution of Marek's disease in chickens.

Witte (1998), Chang et al. (2015), Jørgensen et al. (2016), and Hiltunen et al. (2017) discuss antibiotic resistance.

The United States in 2017 implemented a voluntary plan with industry to phase out the use of certain antibiotics for enhanced food production and requires the use of antibiotics by prescription only; see https://www.fda.gov/for consumers/consumerupdates/ucm378100.htm. Despite this and other bans in the EU, the level of antibiotic-resistant bacteria in the farms has not gone down to zero. For details, see https://www.danmap.org/downloads/reports.

McKenna (2019) discusses the use of antibiotics in citrus orchards.

Witte (1998) discusses vancomycin resistance.

Roberts reports on our work in Roberts et al. (2016).

In the United States annually there are 19 people per 100,000 infected with *Campylobacter* and 16 people per 100,000 infected with *Salmonella*. For details, see https://foodpoisoningbulletin.com/2018/mmwr-examines-trends-food borne-illness-infections-2006-2017/.

Chapter 13

Springmann et al. (2018) discuss planetary limits and how our agricultural system can be kept within them.

Bennett et al. (2016) note that major shifts in societal behavior emerge in the face of major social-environmental challenges.

Springmann et al. (2018) project environmental conditions to 2050 and 2100 that assess changing diet, reducing waste, and closing yield gaps in relation to continuing the status quo. These scientists assess change in environmental strain when considering each change individually and collectively, and do so for moderate and substantial improvements relative to the status quo.

Godfray et al. (2010), Foley (2011), Foley et al. (2011), Tilman et al. (2011), Tilman and Clark (2015), Robertson (2015), Mueller and Binder (2015), Jahn et al. (2015), Erb et al. (2016), and Springmann et al. (2018) review the importance of changing diet, reducing yield gaps, and reducing waste to future agriculture.

Koneswaran and Nierenberg (2008), Weber and Matthews (2008), Wirsenius et al. (2010), Foley (2011), and Eisler et al. (2014) discuss the advantages of grass-raised beef and ways to reduce cattle production of greenhouse gas. Hristov (2012) considers the contribution of bison and other wild animals to greenhouse gas emissions.

Heffernan (2017) and Milius (2018) report on new plant-based alternatives to meat.

Erb et al. (2016), Springmann et al. (2018), and Lymbery (2018) extoll the advantages of a flexitarian diet. Tilman and Clark (2014) and Jahn et al. (2015) consider the ecological and health benefits of such diets.

Stokstad (2010) and Durisin and Singh (2018) analyze the United States' consumption of beef and meat.

Koneswaran and Nierenberg (2008), Cooke et al. (2008), Matson et al. (2013), and Zumkehr and Campbell (2015) discuss food hubs and the value of buying local.

Weber and Matthews (2008) quantify the greenhouse gas implications of local purchases versus moving toward a plant-based diet.

Saha (2017a,b) discusses environmental certification.

Information about Joel Gamoran can be found at http://www.joelgamoran.com/.

Byerly et al. (2018) review strategies to change human behavior, including the use of smaller plates to reduce food waste.

A program that uses vending machines to supply surplus food to the homeless is described at https://www.actionhunger.org.

Tilman and Clark (2015) give examples (including the French grocery chain Intermarché) of how encouraging sale of irregular produce or that which is nearing its expiration date reduces food waste. Godfray et al. (2010) discuss the challenges to reduce food waste in developed versus developing countries.

Mueller et al. (2012), Mueller and Binder (2015), and Springmann et al. (2018) consider yield gaps and strategies to close them.

Pennisi (2010), Tester and Langridge (2010), Fedoroff et al. (2010), Bevan et al. (2017), and Godfray et al. (2018) review genetic modification of plants to improve their yields and abilities to tolerate environmental stress and reduced fertilizer treatments.

Gilbert (2016) reports on traditional breeding programs that increase crop yield, e.g., bean production.

Gebbers and Adamchuk (2010) and King (2017) review the use of sensors and other high-tech devices to improve farming.

Mueller et al. (2012) and Zhang (2017) discuss efficient distribution of fertilizers to increase yield and reduce pollution.

Roberts (2018) reports on Duke Energy's program.

Bray (2017) reports on Jeremy Visser's dairy project.

Raine (2018) and Siviter et al. (2018) discuss the problems with neonicotinoid- and sulfoximine-based insecticides.

Délye et al. (2013) and Borel (2017) discuss new approaches to genetically modifying crops to resist disease, selecting microbes useful as pesticides, and the limits of technology to solve the problems with farm pests and weeds. "Technology won't save the farm" is quoted from Borel (2017).

Kemeny (2018) reports on the potential of mint to limit crop damage from insects and rusts.

Zimmerer (2010) reviews the potential of climate change to affect agriculture. Estimated losses to Nebraska farmers and ranchers from the March 13–29 flood event is reported by NPR, https://www.npr.org/2019/03/21/705408364 /nebraska-faces-over-1-3-billion-in-flood-losses.

Meldrum et al. (2018) study the response of Bolivian farmers to climate change. Ancient potatoes from the American Southwest also survived climatic extremes of the past and may offer solutions for the future (Kapoor 2017).

Robertson (2015) argues that the lack of economic incentives is a central barrier to farmers' adoption of sustainable practices. He notes that the political will to incentivize change is needed to overcome some of the most pernicious problems that agriculture poses to the environment.

Agricultural subsidies by country are provided at Wikipedia, s.v. "agricultural subsidy." Details on the subsidies provided in the United States and the EU are from McFadden and Hoppe (2017) and https://ec.europa.eu/info/food -farming-fisheries/key-policies/common-agricultural-policy/cap-glance_en.

Waldron et al. (2017) investigate the relation between investing in conservation and the ability of countries to forestall extinction and reduce the expansion of agriculture.

Stuart et al. (2010) review environmental markets.

Hill et al. (2019) estimate the health and climate change costs of production in the United States.

A variety of programs aim to help young farmers get established, including the
USDA's Transition Incentives Program (https://www.fsa.usda.gov/online
-services/haynet-tipnet/tipnet/index), the International Farm Transition
Network (https://www.farmtransition.org/), and Land Link (https://www
.cfra.org/renewrural/landlink).

Crist et al. (2017), Ejeta (2010), Wright et al. (2012), Fischer et al. (2014), and Liao
and Brown (2018) discuss the importance of land ownership, human well-
being, education, and family planning for women to agricultural sustainability.

Literature Cited

Agnvall, B., Katajamaa, R., Altimiris, J., and P. Jensen. 2015. Is domestication driven by reduced fear of humans? Boldness, metabolism and serotonin levels in divergently selected red junglefowl (*Gallus gallus*). *Biology Letters* 11:20150509.

Allan, B. F., Tallis, H., Chaplin-Kramer, R., Huckett, S., Kowal, V. A., Musengezi, J., Okanga, S., Ostfeld, R. S., Schieltz, J., Warui, C. M., Wood, S. A., and F. Keesing. 2017. Can integrating wildlife and livestock enhance ecosystem services in central Kenya? *Frontiers in Ecology and the Environment* 15: 328–335.

Arendt, M., Cairns, K. M., Ballard, J. W. O., Savolainen, P., and E. Axelsson. 2016. Diet adaptation in dog reflects spread of prehistoric agriculture. *Heredity* 117:301–306.

Askins, R. A., Chávez-Ramírez, C., Dale, B. C., Haas, C. A., Herkert, J. R., Knopf, F. L., and P. D. Vickery. 2007. *Conservation of Grassland Birds in North America: Understanding Ecological Processes in Different Regions. Ornithological Monographs* 64. The American Ornithologists' Union, Washington, D.C.

Atkins, K. E., Read, A. F., Savill, N. J., Renz, K. G., Fakhrul Islam, A. F. M., Walkden-Brown, S. W., and M. E. J. Woolhouse. 2012. Vaccination and reduced cohort duration can drive virulence evolution: Marek's disease virus and industrialized agriculture. *Evolution* 67:851–860.

Azhar, B., Lindeman, D. B., Wood, J., Fischer, J., Manning, A., McElhinny, C., and M. Zakaria. 2013. The influence of agricultural system, stand structural complexity and landscape context on foraging birds in oil palm landscapes. *Ibis* 155:297–312.

Balmford, A., Green, R., and B. Phalan. 2012. What conservationists need to know about farming. *Proceedings of the Royal Society, B* 279:2714–2724.

———. 2015. Land for food and land for nature? *Daedalus* 144:57–75.

Barrett, C. B. 2010. Measuring food insecurity. *Science* 327:825–828.

Baulcombe, D. 2010. Reaping the benefits of crop research. *Science* 327:761.

Beason, R. C. 1995. Horned lark (*Eremophila alpestris*), version 2.0. In: *The Birds of North America* (A. F. Poole and F. B. Gill, editors). Cornell Lab of Ornithology, Ithaca, N.Y.

Beedy, E. C., Hamilton, W. J. III, Meese, R. J., Airola, D. A., and P. Pyle. 2018. Tricolored blackbird (*Agelaius tricolor*), version 3.1. In: *The Birds of North America* (P. G. Rodewald, editor). Cornell Lab of Ornithology, Ithaca, N.Y.

Bennett, E. M., Solan, M., Biggs, R., McPhearson, T., Norstrom, A. V., Olsson, P., Pereira, L., Peterson, G. D., Raudsepp-Hearne, C., Biermann, F., Carpenter, S. R., Ellis, E. C., Hichert, T., Galaz, V., Lahsen, M., Mikloreit, M., López, B. M., Nicholas, K. A., Preiser, R., Vince, G., Vervoort, J. M., and J. Xu. 2016. Bright spots: seeds of a good Anthropocene. *Frontiers in Ecology and the Environment* 14:441–448.

Berry, W. 1977. *The Unsettling of America.* Counterpoint Press, Berkeley, Calif.

Betts, M. G., Wolf, C., Ripple, W. J., Phalan, B., Millers, K. A., Duarte, A., Butchart, S. H. M., and T. Levi. 2017. Global forest loss disproportionately erodes biodiversity in intact landscapes. *Nature* 547:441–444.

Bevan, M. W., Uauy, C., Wulff, B. B. H., Zhou, J., Krasileva, K., and M. D. Clark. 2017. Genomic innovation for crop improvement. *Nature* 543:346–354.

Blackman, A., and M. A. Naranjo. 2012. Does eco-certification have environmental benefits? Organic coffee in Costa Rica. *Ecological Economics* 83:58–66.

Blasiak, R. 2011. Ethics and environmentalism: Costa Rica's lesson. December 7. United Nations University Our World 2.0. https://ourworld.unu.edu/en /ethics-and-environmentalism-costa-ricas-lesson.

Borel, B. 2017. When the pesticides run out. *Nature* 543:302–304.

Bowers, J. H., and B. A. Bailey. 2001. The impact of plant diseases on world chocolate production. *Plant Health Progress.* doi:10.1094/PHP-2001-0709-01-RV.

Bray, K. 2017. Coming full circle. *Everett (Wash.) Herald.* May 5.

Browning, M., Cleckler, J., Knott, K., and M. Johnson. 2016. Prey consumption by a large aggregation of barn owls in an agricultural setting. In: *Proceedings of the 27th Vertebrate Pest Conference,* pp. 337–344 (R. M. Timm and R. A. Baldwin, editors). University of California, Davis.

Buij, R., Folkertsma, I., Kortekaas, K., de Iongh, H. H., and J. Komdeur. 2013. Effects of land-use change and rainfall in Sudano-Sahelian West Africa on the diet and nestling growth rates of an avian predator. *Ibis* 155:89–101.

Butler, D. 2018. EU pesticide review could lead to ban. *Nature* 555:150–151.

Byerly, H., Balmford, A., Ferraro, P. J., Wagner, C. H., Palchak, E., Polasky, S., Ricketts, T. H., Schwartz, A. J., and B. Fisher. 2018. Nudging pro-environmental behavior: evidence and opportunities. *Frontiers in Ecology and the Environment* 16:159–168.

Carneiro, M., Afonso, S., Geraldes, A., Garreau, H., Bolet, G., Boucher, S., Tircazes, A., Queney, G., Nachman, M. W., and N. Ferrand. 2011. The genetic structure of domestic rabbits. *Molecular Biology and Evolution* 28:1801–1816.

Carvell, C., Bourke, A. F. G., Dreier, S., Freeman, S. N., Hulmes, S., Jordan, W. C., Redhead, J. W., Sumner, S., Wang, J., and M. S. Heard. 2017. Bumblebee family lineage survival is enhanced in high-quality landscapes. *Nature* 543: 547–549.

Caspermeyer, J. 2017. Holy chickens: did medieval religious rules drive domestic chicken evolution? *Molecular Biology and Evolution* 34:2123–2124.

Castañeda, X. A. 2018. *Hunting habitat use and selection patterns of barn owl (*Tyto alba*) in the urban-agricultural setting of a prominent wine grape growing region of California.* M.Sc. thesis. Humboldt State University.

Chandler, R. B., King, D. I., Raudales, R., Trubey, R., Chandler, C., and V. J. A. Chávez. 2013. A small-scale land-sparing approach to conserving biological diversity in tropical agricultural landscapes. *Conservation Biology* 27: 785–795.

Chang, Q., Wang, W., Regev-Yochay, G., Lipsitch, M., and W. P. Hanage. 2015. Antibiotics in agriculture and the risk to human health: how worried should we be? *Evolutionary Applications* 8:240–245.

Charnley, S., Gosnell, H., Wendel, K. L., Rowland, M. M., and M. J. Wisdom. 2018. Cattle grazing and fish recovery on US federal lands: can social-ecological systems science help? *Frontiers in Ecology and the Environment* 16:S11–S22.

Childs, C. 2012. *Apocalyptic Planet.* Vintage Books, New York.

Choi, J. Y., Platts, A. E., Fuller, D. Q., Hsing, Y.-L., Wing, R. A., and M. D. Purugganan. 2017. The rice paradox: multiple origins but single domestication in Asian rice. *Molecular Biology and Evolution* 34:969–979.

Chomicki, G., and S. S. Renner. 2016. Obligate plant farming by a specialized ant. *Nature Plants* 2:16181.

Clark, S. G., and M. B. Rutherford (editors). 2014. *Large Carnivore Conservation.* University of Chicago Press, Chicago.

Clucas, B., Marzluff, J. M., Mackovjak, D., and I. Plamquist. 2013. Do American crows pay attention to human gaze and facial expression? *Ethology* 119: 296–302.

Conner, R., Seidl, A., VanTasell, L., and N. Wilkins. 2001. *United States Grasslands and Related Resources: An Economic and Biological Trends Assessment.* Unpublished report, available at https://nri.tamu.edu/publications/research-reports/2001/united-states-grasslands-and-related-resources-an-economic-and-biological-trends-assessment/.

Cooke, A. M., Curran, S. R., Linton, A., and A. Schrank. 2008. Introduction: agriculture, trade and the global governance of food. *Globalizations* 5:99–106.

Cossins, D. 2015. Amazing animal farmers that grow their own food. http://www
.bbc.com/earth/story/20150105-animals-that-grow-their-own-food.

Coulson, J. O., Rondeau, E., and M. Caravaca. 2018. Yellow-headed caracara and
black vulture cleaning Baird's tapir. *Journal of Raptor Research* 52:104–107.

Cressey, D. 2017. Neonics vs bees. *Nature* 551:156–158.

Creutzig, F. 2017. Govern land as a global commons. *Nature* 546:28–29.

Crist, E., Mora, C., and R. Engelman. 2017. The interaction of human population,
food production, and biodiversity protection. *Science* 356:260–264.

Crosby, A. D., Elmore, R. D., Leslie, D. M. Jr., and R. E. Will. 2015. Looking beyond rare species as umbrella species: Northern bobwhites (*Colinus virgi-
nianus*) and conservation of grassland and shrubland birds. *Biological Conservation* 186:233–240.

Crowder, D. W., and J. P. Reganold. 2015. Financial competitiveness of organic
agriculture on a global scale. *Proceedings of the National Academy of Sciences
(USA)* 112:7611–7616.

Cui, Z., Hongyan, Z., Chen, X., Zhang, C., Ma, W., Huang, C., Zhang, W.,
Mi, G., Miao, Y., Li, X., Gao, Q., Yang, J., Wang, Z., Ye, Y., Guo, S., Lu, J.,
Huang, J., Lv, S., Sun, Y., Liu, Y., Peng, X., Ren, J., Li, S., Deng, X., Shi, X.,
Zhang, Q., Yang, Z., Tang, L., Wei, C., Jia, L., Zhang, J., He, M., Tong, Y.,
Tang, Q., Zhong, X., Liu, Z., Cao, N., Kou, C., Ying, H., Yin, Y., Jiao, X.,
Zhang, Q., Fan, M., Jiang, R., Zhang, F., and Z. Dou. 2018. Pursuing sus-
tainable productivity with millions of smallholder farmers. *Nature* 555:363–
366.

Cutright, P. R. 1989. *Lewis and Clark: Pioneering Naturalists.* Bison Books, University of Nebraska Press, Lincoln.

Dalin, C., Wada, Y., Kastner, T., and M. J. Puma. 2017. Groundwater depletion embedded in international food trade. *Nature* 543:700–704.

Darawshi, S., Leshem, Y., and U. Motro. 2017. Aggregations and dietary changes
of short-toed snake-eagles: a new phenomenon associated with modern agriculture. *Journal of Raptor Research* 51:446–450.

de Camino, R., Segura, O., Aria, L. G., and I. Pérez. 2000. *Costa Rica: Forest
Strategy and the Evolution of Land Use.* World Bank, Washington, D.C.

Délye, C., Jasieniuk, M., and V. Le Corre. 2013. Deciphering the evolution of herbicide resistance in weeds. *Trends in Genetics* 29:649–658.

de Ponti, T., Rijk, B., and M. K. Van Ittersum. 2012. The crop yield gap between
organic and conventional agriculture. *Agricultural Systems* 108:1–9.

Dewey, C. 2017. Growing number of young people quit desk job to become farmers.
Washington Post. November 23.

Diamond, J. 1993. *The Third Chimpanzee.* Harper Perennial, New York.

Díaz-Pascacio, E., Ortega-Argueta, A., Castillo-Uzcanga, M. M., and N. Ramírez-Marcial. 2018. Influence of land use on the riparian zone condition along an urban-rural gradient on the Sabinal River, Mexico. *Botanical Sciences* 96: 180–199.

Dow, G. K., and C. G. Reed. 2011. Stagnation and innovation before agriculture. *Journal of Economic Behavior and Organization* 77:339–350.

Dross, C., Jiguet, F., and M. Tichit. 2017. Concave trade-off curves between crop production and functional and phylogenetic diversity of birds. *Ecological Indicators* 79:83–90.

Durisin, M., and S. D. Singh. 2018. U.S. set to break meat-eating record. *Bloomberg News.* January 3.

Eisler, M. C., Lee, M. R. F., Tarlton, J. F., Martin, G. B., Beddington, J., Dungait, J. A. J., Greathead, H., Liu, J., Mathew, S., Miller, H., Misselbrook, T., Murray, P., Vinod, V. K., Van Saun, R., and M. Winter. 2014. Steps to sustainable livestock. *Nature* 507:32–34.

Ejeta, G. 2010. African Green Revolution needn't be a mirage. *Science* 327:831–832.

Ellis, E., Maslin, M., Boivin, N., and A. Bauer. 2016. Involve social scientists in defining the Anthropocene. *Nature* 540:192–193.

Erb, K.-H., Lauk, C., Kastner, T., Mayer, A., Theurl, M. C., and H. Haber. 2016. Exploring the biophysical option space for feeding the world without deforestation. *Nature Communications* 7:11382.

Ewing, J. 2005. *Monkeys Are Made of Chocolate.* PixyJack Press, Masonville, Colo.
———. 2015. *Where Tapirs and Jaguars Once Roamed.* PixyJack Press, Masonville, Colo.

Farrell, B. D., Sequeira, A. S., O'Meara, B. C., Normark, B. B., Chung, J. H., and B. H. Jordal. 2001. The evolution of agriculture in beetles (Curculionidae: Scolytinae and Platypodinae). *Evolution* 55:2011–2027.

Fedoroff, N. V., Battisti, D. S., Beachy, R. N., Cooper, P. J. M., Fischhoff, D. A., Hodges, C. N., Knauf, V. C., Lobell, D., Mazur, B. J., Molden, D., Reynolds, M. P., Ronald, P. C., Rosegrant, M. W., Sanchez, P. A., Vonshak, A., and J.-K. Zhu. 2010. Radically rethinking agriculture for the 21st century. *Science* 327:833–834.

Fischer, J., Abson, D. J., Butsic, V., Chappell, M. J., Ekroos, J., Hanspach, J., Kuemmerle, T., Smith, H. G., and H. von Wehrden. 2014. Land sparing versus land sharing: moving forward. *Conservation Letters* 7:149–157.

Fitzmaurice, C. J., and B. J. Gareau. 2016. *Organic Futures.* Yale University Press, New Haven.

Foley, J. A. 2011. Can we feed the world and sustain the planet? *Scientific American* 305:60–65.

———. 2013. It's time to rethink America's corn system. *Scientific American.* March 5. Available at https://www.scientificamerican.com/article/time-to -rethink-corn/.

Foley, J. A., Ramankutty, N., Brauman, K. A., Cassidy, E. S., Gerger, J. S., Johnston, M., Mueller, N. D., O'Connell, C. O., Ray, D. K., West, P. C., Balzer, C., Bennett, E. M., Carpenter, S. R., Hill, J., Monfreda, C., Polasky, S., Rockström, J., Sheehan, J., Siebert, S., Tilman, D., and D. P. M. Zacks. 2011. Solutions for a cultivated planet. *Nature* 478:337–342.

FAO (Food and Agriculture Organization of the United Nations). 2017. *The Future of Food and Agriculture — Trends and Challenges.* FAO, Rome.

Frishkoff, L. O., Karp, D. S., M'Gonigle, L. K., Mendenhall, C. D., Zook, J., Kremen, C., Hadly, E. A., and G. C. Daily. 2014. Loss of avian phylogenetic diversity in neotropical agricultural systems. *Science* 345:1343–1346.

Furumo, P. R., and T. M. Aide. 2017. Characterizing commercial oil palm expansion in Latin America: land use change and trade. *Environmental Research Letters* 12:024008.

Gabriel, D., Sait, S. M., Kunin, W. E., and T. G. Benton. 2013. Food production vs. biodiversity: comparing organic and conventional agriculture. *Journal of Applied Ecology* 50:355–364.

Garfinkel, M., and M. Johnson. 2015. Pest-removal services provided by birds on small organic farms in northern California. *Agriculture, Ecosystems and Environment* 211:24–31.

Gebbers, R., and V. I. Adamchuk. 2010. Precision agriculture and food security. *Science* 237:828–831.

Gilbert, N. 2016. Frugal farming. *Nature* 533:308–310.

Giordano, M. 2009. Global groundwater? Issues and solutions. *Annual Review of Environment and Resources* 34:153–178.

Glausiusz, J. 2018. Owls for peace. *Nature* 554:22–23.

Gleeson, T., Wada, Y., Bierkens, M. F. P., and L. P. H. van Beek. 2012. Water balance of global aquifers revealed by groundwater footprint. *Nature* 488:197–200.

Godfray, H. C., Beddington, J. R., Crute, I. R., Haddad, L., Lawrence, D., Muir, J. F., Pretty, J., Robinson, S., Thomas, S. M., and C. Toulmin. 2010. Food security: the challenge of feeding 9 billion people. *Science* 327:812–818.

González-Fortes, G., Jones, E. R., Lightfoot, E., Bonsall, C., Lazar, C., Grandald'Anglade, A., Garralda, M. D., Drak, L., Siska, V., Simalcsik, A., Boroneant, A., Romaní, J. R. V., Rodríguez, M. V., Arias, P., Pinhasi, R., Manica, A., and M. Hofreiter. 2017. Paleogenomic evidence for multigenerational mixing between Neolithic farmers and Mesolithic huntergatherers in the Lower Danube Basin. *Current Biology* 27:1801–1810.

Goulson, D. 2015. Field study. *Smithsonian Magazine* April, p. 22.

Granatstein, D. 1992. *Dryland Farming in the Northwestern United States: A Nontechnical Review.* MISC0162. Washington State University Cooperative Extension, Pullman.

Green, R. E., Cornell, S. J., Scharlemann, J. P. W., and A. Balmford. 2005. Farming and the fate of wild nature. *Science* 307:550–555.

Guinan, J. A., Gowaty, P. A., and E. K. Eltzroth. 2008. Western bluebird (*Sialia mexicana*), version 2.0. In: *The Birds of North America* (A. F. Poole and F. B. Gill, editors). Cornell Lab of Ornithology, Ithaca, N.Y.

Guzmán, R. A., and J. Weisdorf. 2011. The Neolithic Revolution from a price-theoretic perspective. *Journal of Development Economics* 96:209–219.

Guzy, M. J., and G. Ritchison. 1999. Common yellowthroat (*Geothlypis trichas*), version 2.0. In: *The Birds of North America* (A. F. Poole and F. B. Gill, editors). Cornell Lab of Ornithology, Ithaca, N.Y.

Hamilton, S. K., Doll, J. E., and G. P. Robertson (editors). 2015. *The Ecology of Agricultural Landscapes.* Oxford University Press, New York.

Hanspach, J., Abson, D. J., Collier, N. F., Dorresteijn, I., Schultner, J., and J. Fischer. 2017. From trade-offs to synergies in food security and biodiversity conservation. *Frontiers in Ecology and the Environment* 15:489–494.

Harrison, J. O., Brown, M. B., Powell, L. A., Schacht, W. H., and J. A. Smith. 2017. Nest site selection and nest survival of greater prairie-chickens near a wind energy facility. *Condor* 119:659–672.

Hass, A. L., Kormann, U. G., Tscharntke, T., Clough, Y., Baillod, A. B., Sirami, C., Fahrig, L., Martin, J.-L., Baudry, J., Bertrand, C., Bosch, J., Brotons, L., Burel, F., Georges, R., Giralt, D., Marcos-García, M. A., Ricarte, A., Siriwardena, G., and P. Batáry. 2018. Landscape configurational heterogeneity by small-scale agriculture, not crop diversity, maintains pollinators and plant reproduction in western Europe. *Proceedings of the Royal Society, B* 285:20172242.

Hausen, J. 2013. MSU honors Belgrade ranchers. *Bozeman (Mont.) Daily Chronicle.* October 26.

Heath, S. K., Soykan, C. U., Velas, K. L., Kelsey, R., and S. M. Kross. 2017. A bustle in the hedgerow: woody field margins boost on farm avian diversity and abundance in an intensive agricultural landscape. *Biological Conservation* 212:153–161.

Hébert, M.-P., Fugère, V., and A. Gonzalez. 2019. The overlooked impact of rising glyphosate use on phosphorous loading in agricultural watersheds. *Frontiers in Ecology and the Environment* 17:48–56.

Heffernan, O. 2017. A meaty issue. *Nature* 544:S18–S20.

Henn, B. M., Cavalli-Sforza, L. L., and M. W. Feldman. 2012. The great human expansion. *Proceedings of the National Academy of Sciences (USA)* 109:17758–17764.

Henning, B. G. 2015. The ethics of food, fuel and feed. *Daedalus* 144:990–998.

Herkert, J. R. 2009. Response of bird populations to farmland set-aside programs. *Conservation Biology* 23:1036–1040.

Hilborn, R., Banobi, J., Hall, S. J., Pucylowski, T., and T. E. Walsworth. 2018. The environmental cost of animal source foods. *Frontiers in Ecology and the Environment* 16:329–335.

Hill, J., Goodkind, A., Tessum, C., Thakrar, S., Tilman, D., Polasky, S., Smith, T., Hunt, N., Mullins, K., Clark, M., and J. Marshall. 2019. Air-quality-related health damages of maize. *Nature Sustainability* 2:397–403.

Hiltunen, T., Virta, M., and A.-L. Laine. 2017. Antibiotic resistance in the wild: an eco-evolutionary perspective. *Philosophical Transactions of the Royal Society, B* 372:20160039.

Hoelle, J. 2012. Black hats and smooth hands: elite status, environmentalism, and work among the ranchers of Acre, Brazil. *Anthropology of Work Review* 33:60–72.

Howard, K. A., and M. D. Johnson. 2014. Effects of natural habitat on pest control in California vineyards. *Western Birds* 45:276–283.

Hristov, A. N. 2012. Historic, pre-European settlement, and present-day contribution of wild ruminants to enteric methane emissions in the United States. *Journal of Animal Science* 90:1371–1375.

Hulme, M. F., Vickery, J. A., Green, R. E., Phalan, B., Chamberlain, D. E., Pomeroy, D. E., Nalwanga, D., Mushabe, D., Katebaka, R., Bolwig, S., and P. W. Atkinson. 2013. Conserving the birds of Uganda's banana-coffee arc: land sparing and land sharing compared. *PLoS ONE* 8:e54597.

Jahn, J. L., Stampfer, M. J., and W. C. Willett. 2015. Food, health and the environment: a grand global challenge and some solutions. *Daedalus* 144:31–44.

Jedlicka, J. A., Greenberg, R., and D. K. Letourneau. 2011. Avian conservation practices strengthen ecosystem services in California vineyards. *PLoS ONE* 6:e27347.

Johnson, D. H., and L. D. Igl. 1995. Contributions of the Conservation Reserve Program to populations of breeding birds in North Dakota. *Wilson Bulletin* 107:709–718.

Jørgensen, P. S., Wernli, D., Carroll, S. P., Dunn, R. R., Harbarth, S., Levin, S. L., So, A. D., Schlüter, M., and R. Laxminarayan. 2016. Use antimicrobials wisely. *Nature* 537:159–161.

Kaiser-Bunbury, C. N., Mougal, J., Whittington, A. E., Valentin, T., Gabriel, R., Olesen, J. M., and N. Blüthgen. 2017. Ecosystem restoration strengthens pollination network resilience and function. *Nature* 542:223–227.

Kapoor, M. L. 2017. Lessons from the Holocene. *High Country News.* November 27, p. 5.

Karp, D. S., Gennet, S., Kilonzo, C., Partyka, M., Chaumont, N., Atwill, E. R., and C. Kremen. 2015. Comanaging fresh produce for nature conservation and food safety. *Proceedings of the National Academy of Sciences (USA)* 112: 11126–11131.

Kellermann, J. L., Johnson, M. D., Stercho, A. M., and S. C. Hackett. 2007. Ecological and economic services provided by birds on Jamaican Blue Mountain coffee farms. *Conservation Biology* 22:1177–1185.

Kemeny, R. 2018. Planting mint may reduce crop insecticide use. *Frontiers in Ecology and the Environment* 16:437.

King, A. 2017. The future of agriculture. *Nature* 544:S21–S23.

Knopf, F. L. 1994. Avian assemblages on altered grasslands. *Studies in Avian Biology* 15:247–257.

Kohler, T. A., Smith, M. E., Bogaard, A., Feinman, G. M., Peterson, C. E., Betzenhauser, A., Pailes, M., Stone, E. C., Prentiss, A. M., Dennehy, T. J., Ellyson, L. J., Nicholas, L. M., Faulseit, R. K., Styring, A., Whitlam, J., Fochesato, M., Foor, T. A., and S. Bowles. 2017. Greater post-Neolithic wealth disparities in Eurasia than in North America and Mesoamerica. *Nature* 551: 619–622.

Komar, O. 2006. Ecology and conservation of birds in coffee plantations: a critical review. *Bird Conservation International* 16:1–23.

Koneswaran, G., and D. Nierenberg. 2008. Global farm animal production and global warming: impacting and mitigating climate change. *Environmental Health Perspectives* 116:578–582.

Kremen, C., Iles, A., and C. Bacon. 2012. Diversified farming systems: an agroecological, systems-based alternative to modern industrial agriculture. *Ecology and Society* 17:44. http://dx.doi.org/10.5751/ES-05103-170444.

Kross, S. M. 2012. *The Efficacy of Reintroducing the New Zealand Falcon into the Vineyards of Marlborough for Pest Control and Falcon Conservation.* Ph.D. diss. University of Canterbury, Christchurch, New Zealand.

Kross, S. M., Bourbour, R. P., and B. L. Martinico. 2016. Agricultural land use, barn owl diet, and vertebrate pest control implications. *Agriculture, Ecosystems and Environment* 223:167–174.

Kross, S. M., Tylianakis, J. M., and X. J. Nelson. 2011. Effects of introducing threatened falcons into vineyards on abundance of Passeriformes and bird damage to grapes. *Conservation Biology* 26:142–149.

———. 2012. Translocation of New Zealand falcons to vineyards increases nest attendance, brooding and feeding rates. *PLoS ONE* 7:e38679.

———. 2013. Diet composition and prey choice of New Zealand falcons nesting in anthropogenic and natural habitats. *New Zealand Journal of Ecology* 37: 51–59.

Kross, S. M., Kelsey, T. R., McColl, C. J., and J. M. Townsend. 2016. Field-scale habitat complexity enhances avian conservation and avian-mediated pest-control services in an intensive agricultural crop. *Agriculture, Ecosystems and Environment* 225:140–149.

Lacy, A. E., Barzen, J. A., Moore, D. M., and K. E. Norris. 2015. Changes in the number and distribution of greater sandhill cranes in the eastern population. *Journal of Field Ornithology* 86:317–325.

Larson, G., and D. Q. Fuller. 2014. The evolution of animal domestication. *Annual Review of Ecology, Evolution, and Systematics* 45:115–136.

Lazaridis, I., Nadel, D., Rollefson, G., Merrett, D. C., Rohland, N., Mallick, S., Fernandes, D., Novak, M., Gamarra, B., Sirak, K., Connell, S., Steward-son, K., Harney, E., Fu, Q., Gonzalez-Fortes, G., Jones, E. R., Roodenberg, S. A., Lengyel, G., Bocquentin, F., Gasparian, B., Monge, J. M., Gregg, M., Eshed, V., Mizrahi, A-S., Meiklejohn, C., Gerritsen, F., Bejenaru, L., Blü-her, M., Campbell, A., Cavalleri, G., Comas, D., Frogel, P., Gilbert, E., Kerr, S. M., Kovacs, P., Krause, J., McGettigan, D., Merrigan, M., Merriwether, D. A., O'Reilly, S., Richards, M. B., Semino, O., Shamoon-Pour, M., Ste-fanescu, G., Stumvoll, M., Tönjes, A., Torroni, A., Wilson, J. F., Yengo, L., Hovhannisyan, N. A., Patterson, N., Pinhasi, R., and D. Reich. 2016. Genomic insights into the origin of farming in the ancient Near East. *Nature* 536:419–424.

Lebel, L., Lebel, P., Garden, P., Giap, D. H., Khrutmuang, S., and S. Nakayama. 2008. Places, chains, and plates: governing transitions in the shrimp aqua-culture production-consumption system. *Globalizations* 5:211–226.

Leck, C. F. 1979. Avian extinctions in an isolated tropical wet-forest preserve, Ecuador. *Auk* 96:343–352.

Leopold, A. 1931. Game restoration by cooperation on Wisconsin farms. *Wisconsin Agriculturist and Farmer.* April 18.

———. 1939a. Wild feeds on farms. *Wisconsin Agriculturist and Farmer.* November 18.

———. 1939b. Farmer as conservationist. Reprinted and discussed in S. L. Flader and J. B. Callicott, 1991, *The River of the Mother of God.* University of Wisconsin Press, Madison.

———. 1949. *A Sand County Almanac.* Oxford University Press, New York.

Liao, C., and D. G. Brown. 2018. Assessments of synergistic outcomes from sus-tainable intensification of agriculture need to include smallholder liveli-hoods with food production and ecosystem services. *Current Opinion in Environmental Sustainability* 32:53–59.

Lichtenberg, E. M., Kennedy, C. M., Kremen, C., Batáry, P., Berendse, F., Bommarco, R., Bosque-Pérez, N. A., Carvalheiro, L. G., Snyder, W. E., Williams, N. M., Winfree, R., Klatt, B. K., Åström, S., Benjamin, F., Brittain, C., Chaplin-Kramer, R., Clough, Y., Danforth, B., Diekötter, T., Eigenbrode, S. D., Ekroos, J., Elle, E., Freitas, B. M., Fukuda, Y., Gaines-Day, H. R., Grab, H., Gratton, C., Holzschuh, A., Isaacs, R., Isaia, M., Jha, S., Jonason, D., Jones, V. P., Klein, A.-M., Krauss, J., Letourneau, D. K., Mcfadyen, S., Mallinger, R. E., Martin, E. A., Martinez, E., Memmott, J., Morandin, L., Neame, L., Otieno, M., Park, M. G., Pfiffner, L., Pocock, M. J. O., Ponce, C., Potts, S. G., Poveda, K., Ramos, M., Rosenheim, J. A., Rundlöf, M., Sardiñas, H., Saunders, M. E., Schon, N. L., Sciligo, A. R., Sidhu, C. S., Steffan-Dewenter, I., Tschamtke, T., Vesely, M., Weisser, W. W., Wilson, J. K., and D. W. Crowder. 2017. A global synthesis of the effects of diversified farming systems on arthropod diversity within fields and across agricultural landscapes. *Global Change Biology* 23:4946-4957.

Lipinski, A., Limb, R., Sedivec, K., and B. Geaumont. 2014. Grassland bird, rangeland vegetation and prairie dogs on grazed mixed-grass prairie. *North Dakota Beef Report* 2014:49-51.

Loog, L., Thomas, M. G., Barnett, R., Allen, R., Sykes, N., Paxinos, P. D., Lebrasseur, P., Dobney, K., Peters, J., Manica, A., Larson, G., and A. Eriksson. 2017. Inferring allele frequency trajectories from ancient DNA indicates that selection on a chicken gene coincided with changes in medieval husbandry practices. *Molecular Biology and Evolution* 34:1981-1990.

Lymbery, P. 2018. Promote flexitarian diets worldwide. *Nature* 563:325.

MacDonald, A. L. 2012. *Blurring the lines between production and conservation lands: bird use of prairie strips in row-cropped landscapes.* M.Sc. thesis, Iowa State University. Available at http://lib.dr.iastate.edu/etd/12771.

Malan, J. C., Flint, N., Jackson, E. L., Irving, A. D., and D. L. Swain. 2018. Offstream watering points for cattle: protecting riparian ecosystems and improving water quality? *Agriculture, Ecosystems and Environment* 256:144-152.

Mandema, F. S., Tinbergen, J. M., Ens, B. J., Koffijberg, K., Dijkema, K. S., and J. P. Bakker. 2015. Moderate livestock grazing of salt, and brackish marshes benefits breeding birds along the mainland coast of the Wadden Sea. *Wilson Journal of Ornithology* 127:467-476.

Mannion, A. M. 1999. Domestication and the origins of agriculture: an appraisal. *Progress in Physical Geography* 23:37-56.

Manuel, M. de, Kuhlwilm, M., Frandsen, P., Sousa, V. C., Desai, T., Prado-Martinez, J., Hernandez-Rodriquez, J., Dupanloup, I., Lao, O., Hallast, P., Schmidt, J. M., Heredia-Genestar, J. M., Benazzo, A., Barbujani, G., Peter,

B. M., Kuderna, L. F. K., Casals, F., Angedakin, S., Arandjelovic, M., Boesch, C., Kühl, H., Vigilant, L., Langergraber, K., Novembre, J., Gut, M., Gut, I., Navarro, A., Carlsen, F., Andrés, A. M., Siegismund, H. R., Scally, A., Excoffier, L., Tyler-Smith, C., Castellano, S., Xue, Y., Hvilsom, C., and T. Marques-Bonet. 2016. Chimpanzee genomic diversity reveals ancient admixture with bonobos. *Science* 354:477–481.

Marti, C. D., Poole, A. F., and L. R. Bevier. 2005. Barn owl (*Tyto alba*), version 2.0. In: *The Birds of North America* (A. F. Poole and F. B. Gill, editors). Cornell Lab of Ornithology, Ithaca, N.Y.

Marzluff, J. M. 2014. *Welcome to Subirdia.* Yale University Press, New Haven.

Marzluff, J. M., and T. Angell. 2005. *In the Company of Crows and Ravens.* Yale University Press, New Haven.

Marzluff, J. M., and M. Restani. 2018. Extinction and endangerment. In: *Ornithology: Foundation, Critique, and Application,* pp. 779–800 (Morrison, M. L., Rodewald, A. D., Voelker, G., Colón, M. R., and J. F. Prather, editors). Johns Hopkins University Press, Baltimore, Maryland.

Mason, N. A., and P. Unitt. 2018. Rapid change in a native bird population following conversion of the Colorado Desert to agriculture. *Journal of Avian Biology* 2018:e01507.

Mather, M. H., and R. J. Robertson. 1992. Honest advertisement in flight displays of bobolinks (*Dolichonyx oryzivorus*). *Auk* 109:869–873.

Matson, J., Sullins, M., and C. Cook. 2013. *The role of food hubs in local food marketing.* Service Report 73. USDA Rural Development.

Maxwell, S. L., Fuller, R. A., Brooks, T. M., and J. E. M. Watson. 2016. The ravages of guns, nets and bulldozers. *Nature* 536:143–145.

McDermott, A. 2016. Farming spread from two groups. *Science News.* August 6.

McFadden, J. R., and R. A. Hoppe. 2017. *Evolving distribution of payments from commodity, conservation, and federal crop insurance programs.* EIB-184. USDA Economic Research Service.

McGowan, S. 2016. Muddy messages about American migration. *Nature* 537:43–44.

McKenna, M. 2019. Antibiotics hit the orchard. *Nature* 567:302–303.

Meldrum, G., Mijatovic, D., Rojas, W., Flores, J., Pinto, M., Mamani, G., Condori, E., Hilaquita, D., Gruber, H., and S. Padulosi. 2018. Climate change and crop diversity: farmers and adaptation on the Bolivian Altiplano. *Environment, Development and Sustainability* 20:703–730.

Merrill, M. D., Chapman, K. A., Poiani, K. A., and B. Winter. 1999. Land-use patterns surrounding greater prairie-chicken leks in northwestern Minnesota. *Journal of Wildlife Management* 63:189–198.

Meyer, R. S., and M. D. Purugganan. 2016. Evolution of crop species: genetics of domestication and diversification. *Nature Reviews Genetics* 14:840–852.

Milius, S. 2018. Dreaming up tomorrow's burger. *Science News.* September 29, pp. 22–26.

Miller, J., Curtis, T., Chanasyk, D., and W. Willms. 2018. Influence of riparian grazing on channel morphology and riparian health of the Lower Little Bow River. *Canadian Water Resources Journal* 43:18–32.

Milligan, M. C., Johnson, M. D., Garfinkel, M., Smith, C. J., and P. Njoroge. 2016. Quantifying pest control services by birds and ants in Kenyan coffee farms. *Biological Conservation* 194:58–65.

Montgomery, D. R. 2017. *Growing a Revolution.* W. W. Norton, New York.

Morandin, L. A., Long, R. F., and C. Kremen. 2014. Hedgerows enhance beneficial insects on adjacent tomato fields in an intensive agricultural landscape. *Agriculture, Ecosystems and Environment* 189:164–170.

———. 2016. Pest control and pollination cost-benefit analysis of hedgerow restoration in a simplified agricultural landscape. *Journal of Economic Entomology* 109:1020–1027.

Morelli, F. 2018. High nature value farmland increases taxonomic diversity, functional richness and evolutionary uniqueness of bird communities. *Ecological Indicators* 90:540–546.

Morgan, A. M. 2006. Avian influenza: an agricultural perspective. *Journal of Infectious Diseases* 194:S139–S146.

Mueller, H. 1999. Wilson's snipe (*Gallinago delicate*), version 2.0. In: *The Birds of North America* (A. F. Poole and F. B. Gill, editors). Cornell Lab of Ornithology, Ithaca, N.Y.

Mueller, N. D., and S. Binder. 2015. Closing yield gaps: consequences for the global food supply, environmental quality and food security. *Daedalus* 144:45–56.

Mueller, N. D., Gerber, J. S., Johnston, M., Ray, D. K., Ramankutty, N., and J. A. Foley. 2012. Closing yield gaps through nutrient and water management. *Nature* 490:254–257.

Mueller, U. G., Gerardo, N. M., Aanen, D. K., Six, D. L., and T. R. Shultz. 2005. The evolution of agriculture in insects. *Annual Review of Ecology, Evolution, and Systematics* 36:563–595.

Niemeyer, C. 2010. *Wolfer.* Bottlefly Press, Boise, Ida.

Niemuth, N. D. 2005. Landscape composition and great prairie-chicken lek attendance: implications of management. *Prairie Naturalist* 37:127–142.

Niemuth, N. D., Estey, M. E., Fields, S. P., Wangler, B., Bishop, A. A., Moore, P. J., Grosse, R. C., and A. J. Ryba. 2017. Developing spatial models to guide conservation of grassland birds in the U.S. Northern Great Plains. *Condor* 119:506–525.

Palumbi, S. R. 2001. Humans as the world's greatest evolutionary force. *Science* 293:1786–1790.

Patin, E., Laval, G., Barreiro, L. B., Salas, A., Semino, O., Santachiara-Benerecetti, S., Kidd, K. K., Kidd, J. R., Van der Veen, L., Hombert, J.-M., Gessain, A., Froment, A., Bahuchet, S., Heyer, E., and L. Quintana-Murci. 2009. Inferring the demographic history of African farmers and Pygmy hunter-gatherers using a multilocus resequencing data set. *PLoS Genetics* 5:e1000448.

Pearce-Duvet, J. M. C. 2006. The origin of human pathogens: evaluating the role of agriculture and domestic animals in the evolution of human disease. *Biological Reviews* 81:369–382.

Pennisi, E. 2010. Sowing the seeds for the ideal crop. *Science* 327:802–803.

Perfecto, I., and J. Vandermeer. 2010. The agroecological matrix as alternative to the land-sparing/agricultural intensification model. *Proceedings of the National Academy of Sciences (USA)* 107:5786–5791.

Perlut, N. G., and A. M. Strong. 2011. Grassland birds and rotational-grazing in the Northeast: breeding ecology, survival and management opportunities. *Journal of Wildlife Management* 75:715–720.

Perlut, N. G., Strong, A. M., Donovan, T. M., and N. J. Buckley. 2006. Grassland songbirds in a dynamic management landscape: behavioral responses and management strategies. *Ecological Applications* 16:2235–2247.

Pinaria, A. G., Liew, E. C. Y., and L. W. Burgess. 2010. *Fusarium* species associated with vanilla stem rot in Indonesia. *Australasian Plant Pathology* 39:176–183.

Polley, S., Louzada, S., Forni, D., Sironi, M., Balaskas, T., Hains, D. S., Yang, F., and E. J. Hollox. 2015. Evolution of the rapidly mutating human salivary agglutinin gene (DMBT1) and population subsistence strategy. *Proceedings of the National Academy of Sciences (USA)* 112:5105–5110.

Ponisio, L. C., M'Gonigle, L. K., Mace, K. C., Palomino, J., de Valpine, P., and C. Kremen. 2015. Diversification practices reduce organic to conventional yield gap. *Proceedings of the Royal Society, B* 282:20141396.

Pretty, J. N., Noble, A. D., Bossio, D., Dixon, J., Hine, R. E., Penning de Vries, F. W. T., and J. I. L. Morison. 2006. Resource-conserving agriculture increases yields in developing countries. *Environmental Science and Technology* 40:1114–1119.

Prüfer, K., Munch, K., Hellmann, I., Akagi, K., Miller, J. R., Walenz, B., Koren, S., Sutton, G., Kodira, C., Winer, R., Knight, J. R., Mullikin, J. C., Meader, S. J., Ponting, C. P., Lunter, G., Higashino, S., Hobolth, A., Dutheil, J., Karakoç, E., Alkan, C., Sajjadian, S., Catacchio, C. R., Ventura, M., Marques-Bonet, T., Eichler, E. E., André, C., Atencia, R., Mugisha, L., Junhold, J.,

Patterson, N., Siebauer, M., Good, J. M., Fischer, A., Ptak, S. E., Lachmann, M., Symer, D. E., Mailund, T., Schierup, M. H., Andrés, A. M., Kelso, J., and S. Pääbo. 2012. The bonobo genome compared with the chimpanzee and human genomes. *Nature* 486:527-531.

Quinn, J. E., Brandle, J. R., and R. J. Johnson. 2012. The effects of land sparing and wildlife-friendly practices on grassland bird abundance within organic farmlands. *Agriculture, Ecosystems and Environment* 161:10-16.

Railsback, S. F., and M. D. Johnson. 2014. Effects of land use on bird populations and pest control services on coffee farms. *Proceedings of the National Academy of Sciences (USA)* 1211:6109-6114.

Raine, N. E. 2018. A systemic problem with pesticides. *Nature* 561:40-41.

Range, F., Marshall-Pescini, S., Kratz, C., and Z. Virányi. 2019. Wolves lead and dogs follow, but they both cooperate with humans. *Scientific Reports* 9:3796.

Reif, J., Vorisek, P., Stastny, K., Bejcek, V., and P. Jiri. 2008. Agricultural intensification and farmland birds: new insights from a central European country. *Ibis* 150:596-605.

Renfrew, R. B., Hill, J. M., Kim, D. H., Romanek, C., and N. G. Perlut. 2017. Winter diet of bobolink, a long-distance migratory grassland bird, inferred from feather isotopes. *Condor* 119:439-448.

Renfrew, R., Strong, A. M., Perlut, N. G., Martin, S. G., and T. A. Gavin. 2015. Bobolink (*Dolichonyx oryzivorus*), version 2.0. In: *The Birds of North America* (P. G. Rodewald, editor). Cornell Lab of Ornithology, Ithaca, N.Y.

Ribbens, D. 1987. An introduction to the 1947 foreword [to *Great Possessions*]. In: *A Companion to A Sand County Almanac,* pp. 277-288 (J. B. Callicott, editor). University of Wisconsin Press, Madison.

Rice, R. A., and R. Greenberg. 2000. Cacao cultivation and the conservation of biological diversity. *AMBIO* 29:167-173.

Richerson, P. J., Boyd, R., and R. L. Bettinger. 2001. Was agriculture impossible during the Pleistocene but mandatory during the Holocene? A climate change hypothesis. *American Antiquity* 66:387-411.

Richerson, P. J., Boyd, R., and J. Henrich. 2010. Gene-culture coevolution in the age of genomics. *Proceedings of the National Academy of Sciences (USA)* 107:8985-8992.

Ripple, W. J., Smith, P., Haberl, H., Montzka, S. A., McAlpine, C., and D. H. Boucher. 2014. Ruminants, climate change and climate policy. *Nature Climate Change* 4:2-5.

Ritchison, G. 1991. The flight songs of common yellowthroats: description and causation. *Condor* 93:12-18.

Riyahi, S., Hammer, Ø., Arbabi, T., Sánchez, A., Roselaar, C. S., Aliabadian, M., and G.-P. Sætre. 2013. Beak and skull shapes of human commensal and non-commensal house sparrows *Passer domesticus*. *BMC Evolutionary Biology* 13:200.

Roberts, D. 2018. Here's why Duke Energy wants to use pig poop in an NC power plant. *Charlotte Observer.* March 30.

Roberts, M. C., No, D. B., Marzluff, J. M., DeLap, J. H., and R. Turner. 2016. Vancomycin resistant *Enterococcus spp.* from crows and their environment in metropolitan Washington State, USA: is there a correlation between VRE positive crows and the environment? *Veterinary Microbiology* 194:48–54.

Robertson, G. P. 2015. A sustainable agriculture? *Daedalus* 144:76–89.

Sætre, G.-P., Riyahi, S., Aliabadian, M., Hermansen, J. S., Hogner, S., Olsson, U., Gonzalez Rojas, M. F., Sæther, S. A., Trier, C. N., and T. O. Elgvin. 2012. Single origin of human commensalism in the house sparrow. *Journal of Evolutionary Biology* 25:788–796.

Saha, P. 2017a. Beefing up bird habitat. *Audubon* Spring:16.

———. 2017b. Sweet deal. *Audubon* Spring:17.

Salatin, J. 2010. *The Sheer Ecstasy of Being a Lunatic Farmer.* Polyface, Swoope, Va.

Savory, A. 1988. *Holistic Resource Management.* Island Press, Washington, D.C.

Scott, J. C. 2017. *Against the Grain.* Yale University Press, New Haven.

Seto, K. C., and N. Ramankutty. 2016. Hidden linkages between urbanization and food systems. *Science* 352:943–945.

Shaner, D. L. 2014. Lessons learned from the history of herbicide resistance. *Weed Science* 62:427–431.

Shepard, M. 2013. *Restoration Agriculture.* ACRES, Austin, Tex.

Shrubb, M. 2003. *Birds, Scythes and Combines.* Cambridge University Press, New York.

Shryock, B., Marzluff, J. M., and L. M. Moskal. 2017. Urbanization alters the influence of weather and an index of forest productivity on avian community richness and guild abundance in the Seattle metropolitan area. *Frontiers in Ecology and Evolution* 5:40. doi: 10.3389/fevo.2017.00040.

Simonsen, L., Spreeuwenberg, P., Lustig, R., Taylor, R. J., Fleming, D. M., Kroneman, M., Van Kerkhove, M. D., Mounts, A. W., Paget, W. J., and the GLaMOR Colloborating Teams. 2013. Global mortality estimates from the 2009 influenza pandemic from the GLaMOR project: a modeling study. *PLoS Med* 10:e1001558.

Siska, V., Jones, E. P., Jeon, S., Bhak, Y., Kim, H.-M., Cho, Y. S., Kim, H., Lee, K., Veselovskaya, E., Baleuva, T., Gallego-Llorente, M., Hofreiter, M., Bradley, D. G., Eriksson, A., Pinhasi, R., Bhak, J., and A. Manica. 2017. Genome-wide data from two early Neolithic East Asian individuals dating to 7700 years ago. *Science Advances* 3:e1601877.

Siviter, H., Brown, M. J. F., and E. Leadbeater. 2018. Sulfoxaflor exposure reduces bumblebee reproductive success. *Nature* 561:109–112.

Smith, O., Momber, G., Bates, R., Garwood, P., Fitch, S., Pallen, M., Gaffney, V., and R. G. Allaby. 2015a. Sedimentary DNA from a submerged site reveals wheat in the British Isles 8000 years ago. *Science* 347:998–1001.

———. 2015b. Response to comment on Sedimentary DNA from a submerged site reveals wheat in the British Isles 8000 years ago. *Science* 349:247.

Soule, J. D., and J. K. Piper. 1992. *Farming in Nature's Image.* Island Press, Washington, D.C.

Springmann, M., Clark, M., Mason-D'Croz, D., Wiebe, K., Bodirsky, B. L., Lassaletta, L., de Vries, W., Vermeulen, S. J., Herrero, M., Carlson, K. M., Jonell, M., Troell, M., DeClerck, F., Gordon, L. J., Zurayk, R., Scarborough, P., Rayner, M., Loken, B., Franzo, J. Godfray, H. C. J., Tilman, D., Rockström, J., and W. Willett. 2018. Options for keeping the food system within environmental limits. *Nature* 562:519–525.

Stevenson, J. R., Villoria, N., Byerlee, D., Kelley, T., and M. Maredia. 2013. Green Revolution research saved an estimated 18 to 27 million hectares from being brought into agricultural production. *Proceedings of the National Academy of Sciences (USA)* 110:8363–8368.

Stokstad, E. 2010. Could less meat mean more food? *Science* 327:810–811.

Stringer, C., and J. Galway-Witham. 2017. On the origin of our species. *Nature* 546:212–213.

Stuart, D., Canty, D., and K. Killebrew. 2010. *Guide to Environmental Markets for Farmers and Ranchers.* American Farmland Trust, Seattle.

Svedarsky, W. D., Westemeier, R. L., Robel, R. J., Gough, S., and J. E. Toepher. 2000. Status and management of the greater prairie-chicken *Tympanuchus cupido pinnatus* in North America. *Wildlife Biology* 6:277–284.

Swift, K. N., and J. M. Marzluff. 2015. Wild American crows gather around their dead to learn about danger. *Animal Behaviour* 109:187–197.

Tattersall, I., and R. DeSalle. 2015. *A Natural History of Wine.* Yale University Press, New Haven.

Teillard, F., Jiguet, F., and M. Tichit. 2015. The response of farmland bird communities to agricultural intensity as influenced by its spatial aggregation. *PLoS ONE* 10:e0119674.

Teitelbaum, C. S., Converse, S. J., Fagan, W. F., Böhning-Gaese, K., O'Hara, R. B., Lacy, A. E., and T. Mueller. 2016. Experience drives innovation of new migration patterns of whooping cranes in response to global change. *Nature Communications* 7:12793.

Temple, S. A., Fevold, B. M., Paine, L. K., Undersander, D. J., and D. W. Sample.

1999. Nesting birds and grazing cattle: accommodating both on midwestern pastures. *Studies in Avian Biology* 19:178–186.

Tester, M., and P. Langridge. 2010. Breeding technologies to increase crop production in a changing world. *Science* 327:818–822.

Thomas, C. D. 2015. Rapid acceleration of plant speciation during the Anthropocene. *Trends in Ecology and Evolution* 30:448–455.

Thurin, R. 1986. *100 Years of Progress: A History of the First 100 years of Ong, Nebraska*. Privately published.

Tilman, D., and M. Clark. 2015. Food, agriculture and the environment: can we feed the world and save the Earth? *Daedalus* 144:8–23.

Tilman, D., and M. Clark. 2014. Global diets link environmental sustainability and human health. *Nature* 515:518–522.

Tilman, D., Balzer, C., Hill, J., and B. L. Befort. 2011. Global food demand and the sustainable intensification of agriculture. *Proceedings of the National Academy of Sciences (USA)* 108:20260–20264.

Tilman, D., Cassman, K. G., Matson, P. A., Naylor, R., and S. Polasky. 2002. Agricultural sustainability and intensive production practices. *Nature* 418:671–677.

Topál, J., Gásci, M., Miklósi, A., Virányi, Z., Kubinyi, E., and V. Csányi. 2005. Attachment to humans: a comparative study on hand-reared wolves and differently socialized dog puppies. *Animal Behaviour* 70:1367–1375.

Treves, A., Krofel, M., and J. McManus. 2016. Predator control should not be a shot in the dark. *Frontiers in Ecology and the Environment* 14:380–388.

Tucci, S., and J. M. Akey. 2016. A map of human wanderlust. *Nature* 538:179–180.

Tucker, M. 2013. Last days on the ranch. *Belgrade (Mont.) News.* December 27.

Turcotte, M. M., Araki, H., Karp, D. S., Poveda, K., and S. R. Whitehead. 2017. The eco-evolutionary impacts of domestication and agricultural practices on wild species. *Philosophical Transactions of the Royal Society, B* 372: 20160033.

United States Fish and Wildlife Service (USFWS). 2005. *Recovery Plan for Vernal Pool Ecosystems of California and Southern Oregon*. USFWS Region 1. Portland, Oreg.

Varki, A., and T. K. Altheide. 2005. Comparing the human and chimpanzee genomes: search for needles in a haystack. *Genome Research* 15:1746–1758.

Vickery, P. D., and J. R. Herkert (editors). 1999. *Ecology and Conservation of Grassland Birds of the Western Hemisphere*. Studies in Avian Biology 19. Cooper Ornithological Society, Los Angeles.

Vigueira, C. C., Olsen, K. M., and A. L. Caicedo. 2013. The red queen in the corn: agricultural weeds as models of rapid adaptive evolution. *Heredity* 110:303–311.

Vijay, V., Pimm, S. L., Jenkins, C. N., and S. J. Smith. 2016. The impacts of oil palm on recent deforestation and biodiversity loss. *PLoS ONE* 11:e0159668.

Waldron, A., Miller, D. C., Redding, D., Mooers, A., Kuhn, T. S., Nibbelink, N., Timmons Roberts, J., Tobias, J. A., and J. L. Gittleman. 2017. Reductions in global biodiversity loss predicted from conservation spending. *Nature* 551:364–367.

Walk, J. W., Ward, M. P., Benson, T. J., Deppe, J. L., Lischka, S. A., Bailey, S. D., and J. D. Brawn. 2011. *Illinois Birds: A Century of Change.* Illinois Natural History Survey Special Publication 31.

Watson, J. E. M., and O. Venter. 2017. A global plan for nature conservation. *Nature* 550:48–49.

Weber, C. L., and H. S. Matthews. 2008. Food-miles and the relative climate impacts of food choices in the United States. *Environmental Science and Technology* 42:3508–3513.

Wendt, C. A., and M. D. Johnson. 2017. Multi-scale analysis of barn owl nest box selection on Napa Valley vineyards. *Agriculture, Ecosystems and Environment* 247:75–83.

White, C. 2008. *Revolution on the Range.* Shearwater Books, Washington, D.C.

Whitworth, A., Beirne, C., Flatt, E., Huarcaya, R. P., Diaz, J. C. C., Forsyth, A., Molnár, P. K., and J. S. V. Soto. 2018. Secondary forest is utilized by great curassows (*Crax rubra*) and great tinamous (*Tinamus major*) in the absence of hunting. *Condor* 120:852–862.

Wightman, C., Tilly, F., and A. Cilimburg. 2011. *Montana's Colonial-Nesting Waterbird Survey, Final Report.* Montana Fish, Wildlife and Parks, Helena.

Wilkins, A. S., Wrangham, R. W., and W. T. Fitch. 2014. The "domestication syndrome" in mammals: a unified explanation based on neural crest cell behavior and genetics. *Genetics* 197:795–808.

Wilson, J. D., Evans, A. D., and P. V. Grice. 2009. *Bird Conservation and Agriculture.* Cambridge University Press, Cambridge.

Winfree, R., Reilly, J. R., Bartomeus, I., Cariveau, D. P., Williams, N. M., and J. Gibbs. 2018. Species turnover promotes the importance of bee diversity for crop pollination at regional scales. *Science* 359:791–793.

Winkler, D. W. 1996. California gull (*Larus californicus*), version 2.0. In: *The Birds of North America* (A. F. Poole and F. B. Gill, editors). Cornell Lab of Ornithology, Ithaca, N.Y.

Wirsenius, S., Hedenus, F., and K. Mohlin. 2010. Greenhouse gas taxes on animal food products: rationale, tax scheme and climate mitigation effects. *Climatic Change.* doi: 10.1007/s10584-010-9971-x.

Witte, W. 1998. Medical consequences of antibiotic use in agriculture. *Science* 279: 996–997.

Wood, B. 2010. Reconstructing human evolution: achievements, challenges, and opportunities. *Proceedings of the National Academy of Sciences (USA)* 107: 8902–8909.

Woodcock, B. A., Isaac, N. J. B., Bullock, J. M., Roy, D. B., Garthwaite, D. G., Crowe, A., and R. F. Pywell. 2016. Impacts of neonicotinoid use on long-term population changes in wild bees in England. *Nature Communications* 7:12459.

Wright, C. K., and M. C. Wimberly. 2013. Recent land use change in the western corn belt threatens grasslands and wetlands. *Proceedings of the National Academy of Sciences (USA)* 110:4134–4139.

Wright, H. L., Lake, I. R., and P. M. Dolman. 2012. Agriculture—a key element for conservation in the developing world. *Conservation Letters* 5:11–19.

Wuthnow, R. 2011. *Remaking the Heartland.* Princeton University Press, Princeton, N.J.

Yahya, M. S., Syafiq, M., Ashton-Butt, A., Ghazali, A., Asmah, S., and B. Azhar. 2017. Switching from monoculture to polyculture farming benefits birds in oil palm production landscapes: evidence from mist netting data. *Ecology and Evolution* 7:6314–6325.

Yan, W. 2017. A makeover for the world's most hated crop. *Nature* 543:306–308.

Yue, S., Brodie, J. F., Zipkin, E. F., and H. Bernard. 2015. Oil palm plantations fail to support mammal diversity. *Ecological Applications* 25:2285–2292.

Zeder, M. A. 2006. Central questions in the domestication of plants and animals. *Evolutionary Anthropology* 15:105–117.

———. 2011. The origins of agriculture in the Near East. *Current Anthropology* 52:S221–S235.

Zhang, X. 2017. A plan for efficient use of nitrogen fertilizers. *Nature* 543:322–323.

Zheng, Y., Crawford, G. W., Jiang, L., and X. Chen. 2016. Rice domestication revealed by reduced shattering of archaeological rice from the Lower Yangtze Valley. *Scientific Reports* 6:28136.

Zimmerer, K. S. 2010. Biological diversity in agriculture and global change. *Annual Review of Environment and Resources* 35:137–166.

Zuckerman, J. C. 2016. Oil barrens. *Audubon* Fall:24–31.

Zumkehr, A., and J. E. Campbell. 2015. The potential for local croplands to meet US food demand. *Frontiers in Ecology and the Environment* 13:244–248.

Index

Note: Italic page numbers refer to illustrations.